Building the Global Fiber Optics Superhighway

Building the Global Fiber Optics Superhighway

C. David Chaffee
Chaffee Fiber Optics
Ellicott City, Maryland

Kluwer Academic / Plenum Publishers
New York, Boston, Dordrecht, London, Moscow

Library of Congress Cataloging-in-Publication Data

Chaffee, C. David
Building the global fiber optics superhighway/authored by C. David Chaffee.
p. cm.
Includes bibliographical references and index.
ISBN 0-306-46505-1
1. Fiber optics industry. I. Title.

HD9696.F522 C48 2000
338.4′76213692—dc21

00-047985

ISBN 0-306-46505-1

233 Spring Street, New York, New York 10013

http://www.wkap.nl/

10 9 8 7 6 5 4 3 2 1

A C.I.P. record for this book is available from the Library of Congress

Printed in the United States of America

To Katie and Caroline

Acknowledgments

I would like to thank the following for their insight, encouragement, and general support in helping me write this book.

At the top of the list are Kluwer Academic / Plenum editors Tom Cohn and Anna Bozicevic, without whom this book truly would not have been written. Tom, who has since left the company, had the good grace to listen stoically to my rantings about the exploding fiber optics industry one evening in a beverage line at OFC '98, and the rest, as they say, is history. Anna Bozicevic has been a steadying source, whose encouragement has kept the project on course.

Of the many others who also helped immeasurably, I would like to thank Ken Taylor and Larry Johnson for encouraging me to write this book. Both are industry pros who understand the importance of getting the message out. Others who provided expertise include Mike Newsom, Loreli Lees, Cindana Turkatte, Tom Phillips, Steve and Jeff Montgomery, Kevin Tanzillo, Mike Peppler, Duane Piersoll, Paul Rogoski, John Knight, Lawrence Gasman, Loren Talley, Rich Moran, C. David Broecker, Kathleen Coplien Szelag, Pat Robinson, Hans Ehnert, Steve McAbee, Joe Berthold, Barbara Duchez, Mike Unger, Jim Chiddix, Mark Lauroesch, Dave Pangrac, Philip Bell, Julie Unger, Konnie Schaefer, Roger Baker, Garry Adams, Steve Clements, Dr. Don Keck, Cary Bloom, Whit Cotten, John Ryan, James Shaw, Sara Herlihy, Diane Burness, Greg Wortman, Charlie Long, Jennifer Rice, Gordon Lamb, Takashi Touge, Matthew McGuinness, Bill Beck, Mike Mattei, Shelley Grandy, Robie Cline, Peter Westafer, Roger Linscott, Derek Lawrence, Kurt Ruderman, Rachel Woodford, Don Scifres, Andrew Rickman, Mike Chan, Jerry Miller, Jonathan Kraushaar, Jack Kessler, John Pittman, Fred Leonberger, and Dale Niebur.

In the 20 years that I have had the privilege of being involved in the fiber optics industry, many others have shared their time with me to give me a better understanding of the workings of this industry, many of whom are not named here but whose efforts are greatly appreciated.

I would like to thank family members, including my sister-in-law Heide

Lange for sharing her outstanding literary skills and my brother John, her husband, for his continuing support and counsel. I would also like to thank my sister Judi Culbertson for her inspiration and my uncle, Bartlett Hess, now departed, for his encouragement, understanding, and Godly presence.

My daughters, Katie and Caroline, to whom this book is dedicated, are a continuing joy, one that fills my life with love and humor. In them I see a bright hope for tomorrow.

Lastly I wish to thank my parents Captain Hubert Chaffee, now departed, who taught me never to undertake a task without putting everything I have into it to make sure it was done right or not at all, and my mother, Charlotte Chaffee, health now diminished, who has been an unending fount of love, hope, and encouragement throughout my life. May God hold you both in the palm of His hand.

C. David Chaffee

March 2000

Contents

Building the Global Fiber Optics Superhighway

Introduction

Many wonderful stories have contributed to the growth and worldwide renown of the fiber optics industry. From its improbable roots in the 1960s and the important early laser work by Stewart Miller and colleagues at Bell Laboratories to seminal discoveries by Corning's Don Keck, Robert Maurer, and Peter Schultz in 1970 demonstrating that it is feasible to send photons through a glass in a commercially attractive way, fiber optics has been a story of great success and achievement.

In many ways the growth of this technology embodies what is best about our culture and our world. The early discovery and commercialization of fiber optics are a tribute to the free-enterprise system, where creative and ingenious individuals took so many undefined laboratory phenomena and molded them into what has become an absolutely critical communications form for the twenty-first century.

It is relatively easy to recount how fiber optics grew into today's indispensable communications medium. The real work was performed by men and women who took concepts and ideas and engineered them into reality: They are the real heroes of this book. Their work took thousands of man hours and led to many frustrations. Yet in the end it improved and continues to improve how we communicate, do business, indeed even think about the world.

No doubt some of you are familiar with my earlier book, *The Rewiring of America: The Fiber Optics Revolution*, published in 1988. In that book I describe the development of fiber optics from the suggestion of its creation in a paper by Charles Kao and G. A. Hockham in 1966, through its laboratory development, commercial introduction, and marketplace acceptance. Several touchstones from *Rewiring* are important in placing *Building the Global Fiber Optics Superhighway* in perspective. First it is critical to acknowledge the role of AT&T and Corning in the development and growth of fiber optics. Admittedly a different AT&T was involved in the early growth of fiber optics than the divided-up entity we have today, but the work at AT&T and its distinguished Bell Laboratories cannot be overstated. From the first lasers to its development of the fiber, connectors, splitters, and electronics generally—indeed every aspect of this technology—

AT&T defined much of what we now understand as fiber optics. Like an old friend Bell Laboratories continues, although its mission now is split between AT&T and Lucent, among others.

Other telecom laboratories also contributed much, including Bell Northern Research, NTT Labs, BT Labs, and Fujitsu Labs (see Figure I.1).

As important as AT&T/Lucent is Corning, whose understanding of glass led to making the first commercially acceptable optical fiber. Today many of us still marvel at the hair-thin strings that transformed our communications technology and that Corning had the ability to make from a material once considered brittle and difficult to work with. Today using glass to communicate has become commonplace; however 30 years ago it was not, particularly in the minuscule shards that represent an optical fiber. Yet Corning had the vision, the commitment, and the tools to bring it into being.

Figure I.1. Fiber optics has become a highly specialized art. Here a Nortel Networks official studies the characteristics of an optical fiber; results are displayed on a computer screen. (Photo courtesy of Nortel Networks)

Certainly other companies made extremely important contributions to the development of fiber optics, and I chronicle these here as the technology advances at a breathtaking pace. But to understand where advances are occurring, the reader must recall that early technology came largely from these two powerful companies.

In the case of Corning, success involved countless research dollars, a commitment to stay the course in face of a multibillion-dollar concentrated Japanese effort, innovative engineering, a marketplace commitment that led to the construction of a plant before customers, and generally a can-do attitude that experts say saved this extremely important component—optical fiber—for U.S. vendors.

The perceived strength of Corning and AT&T is not an abstract concept. When the technology was first discovered, the two companies engineered a cross-licensing agreement that gave them access to processes and manufacturing technologies that have defined some of fiber's essentials. While at times not so subtle competitors in the marketplace, the two companies have maintained enough of a relationship to update the agreement and keep it intact for some years to come. Upgrading this agreement was important, particularly because of all the intellectual property that has arisen from it. That is probably important, since the original patents that served both Corning and Lucent are expiring.

Intellectual property has been a critical factor in many of Corning's businesses, according to Mark Lauroesch, Corning's division patent counsel for optoelectronics. The company invests heavily in research and development. Corning also defends its patents around the world, particularly as the technology has matured. Lauroesch says:

> A lot of companies are more globally oriented now and it has been important to protect Corning's interests. The company has defended its patents in the United States, Europe and Japan. Such patent litigations are not cheap, however.

Yet Lauroesch acknowledges that Corning is now in a period when many of its original patents are running out. While intellectual property is important, Lauroesch notes that "it is not the only ingredient for success in the fiber industry." There are obviously many accrued benefits from having developed all aspects of the fiber process and the wealth of knowledge that comes from it.

Another success story born from the fiber optics industry is Siecor Corporation, the joint collaboration between Siemens and Corning, which was purchased outright by Corning in late 1999. Siecor was formed in 1977, specifically to develop and market a passive transmission subsystem and cable for the optical fiber that Corning and other vendors were making, according to Derek Lawrence, senior vice-president and general manager of Siecor's cable division. "Corning had developed a fiber that was capable of delivering a transmission medium," says Lawrence in recounting the first days of Siecor's existence. "They had taken optical fiber from scientific exercise to commercial product."

While Corning had some discussions about using AT&T as a cabler, it was clear that AT&T had its own plans to use what was then part of its operation, Western Electric, Lawrence recalls. "The potential market for Corning was going to be through other cable makers," says Lawrence. Corning first went to experienced cable manufacturers in the United States to see if they were interested in teaming but, as Lawrence recalls, "The copper cable manufacturers in the United States didn't seem to embrace the product, perhaps because they didn't have a significant investment in the technology, so they didn't press forward when we presented them with the opportunity."

Siemens, on the other hand, did not have a presence in the United States cable market, and it was interested in obtaining one. The two parties came together and Siecor was born. It became the first nonaffiliated cabler of optical fiber, since Western Electric had done some tests with the AT&T system about the same time. "In terms of developing a commercial product in a noncaptive manner, we were first," says Lawrence.

Siecor emerged strong, and it was able to stay ahead of early competitors. According to Lawrence:

> We had a superior cabling technology compared to the alternative approaches, in which most if not all of our competitors were using a tight buffered design.... It allowed us to cable the fiber with essentially no change in the fiber's properties and no residual stress.... We were able to ensure that this new-fangled stuff would have a reasonable lifetime expectancy.

Siecor's role went far beyond simply selling cable in those days: "We needed to enable the customer," says Lawrence, and "that included splicing, testing, troubleshooting—the broad gamut of field service, including job supervision if that is what the customer wanted." That has been part of the Siecor philosophy, and it remains true today, according to Lawrence. "We see ourselves offering a system solution and not just odd components here and there," he notes. The result is that Siecor has been in the cable accessories market almost from the beginning.

Both of my books focus on how fiber optics is used as a communications medium; that is, how we can better talk to each other, send data, and transmit video using optical fiber. However we should recall that the first applications for fiber optics were not in communications at all but in medicine for imaging in the human body. The possibility of transmitting light throughout the human body to locate potential trouble areas—such as cancerous polyps—using something as small and uninvasive as an optical fiber is not lost on the medical community. In addition, there continues to be ancillary development of optical fiber for noncommunications markets, including display advertising, such as New York City's Times Square; such military applications as deploying missiles tethered to fiber so the operator can see the target before impact, undersea sensing to detect enemy submarines, and helping Air Force planes to fly smarter and lighter.

Multimode fiber was the initial fiber used in fiber optic communications in the United States. However the smaller diameter, capable of longer distances, single-mode fiber soon became the choice for long-distance applications. While some experts thought this would lead to the demise of multimode fiber within a few short years, multimode has found a new niche in office applications, and it is still very much with us today. Part of the continued success of multimode fiber is due to the phenomenal growth of this industry.

At the heart of *Rewiring* was the race among AT&T, Sprint, and MCI to deploy the first coast-to-coast, fiber optics network hookup in 1985. Fiber optics was seen primarily as a means of sending signals long distances, because carriers could avoid the costly and cumbersome repeaters required by long-distance copper, as well as echo, crosstalk, and the other problems inherent in satellite communications.

This nationwide fiber optics race—which was ultimately won by AT&T by a few short weeks—brought the United States fiber optics industry into being in a very real, commercial sense. However in many respects, this early success represented only the first large steps of a toddler. Fiber optics was still in its primitive stages compared to other technologies that had enjoyed decades of growth and maturity. It was not that the use of fiber in long-haul applications lacked importance: After all it had helped to build the best and fastest communications networks in the United States, and other nations were doing the same thing. In addition a brand new market was opening up: Submarine fiber cable was beginning to unite continents.

What frustrated some vendors, and the industry in general, was that once long-haul routes were constructed, they pretty much stopped at the city line. Certainly the Regional Bell Operating Companies (RBOCs) were filling out their networks, and the rise of the first competitive local exchange carriers saw more fiber installation, but there was a major pause following the completion of coast-to-coast routes, which led to slowdowns, lower optical fiber cable production, and somewhat of a brief malaise in the industry. At times trying to support the early infrastructure that arose with the national builds, the fiber optics industry seemed to be a solution in search of a problem.

Yet in these downtimes, creative engineers always seem to do their best—devising novel applications, developing new markets, improving the technology, and finding ways for the industry to flourish again. The capabilities of fiber optics growth is an upward spiral from the time of its origins, including periods of reduced demand. The push to make a better fiber optic system was unrelenting in the world's telecommunications labs and within the companies that had tethered themselves to the prospects of this emerging technology.

The U.S. industry found itself in these crosswinds in 1988, which is where this book begins. Had the fiber optics industry depended on only the U.S. market, there would certainly have been a pronounced downturn. However a far mightier phenomenon was emerging that would compensate for this bump in the road: Other nations had also recognized the advantages of fiber optics and were using it to improve their own networks.

This powerful trend shook the established telecommunications base in countries throughout the globe. After decades of reliance on copper cable and twisted pair, carriers now found communications could be improved by using fiber optics. Networks were being reshaped, relationships recast, and a whole new set of contenders were emerging. This led to dynamic tension between the capabilities of more entrenched suppliers and carriers poised to innovate and adopt fiber optics and new vendors and carriers who claimed they could provide the fresh perspective necessary to grow fully with this amazing technology. That tension—and competition—continues today.

What was taking place was the groundwork for building the global fiber optics superhighway.

Brief Primer[1]

Even the name fiber optics can intimidate nontechnical people who do not know what it means. Optics is simply the study of the properties of light, and fiber refers to the filaments that carry the optics. These are transmitted in massless forms known as photons. Turning that around, an optical fiber is simply a small tube that carries light beams, or lightwaves, as photons.

The idea of optical communications has been with us for many centuries: Native American smoke signals are one primitive example. Optical communications even excited the genius Alexander Graham Bell, who nearly a century prior to the actual implementation of fiber optic systems saw the potential in an experiment he performed in Washington, D.C., on roof tops near the White House.

Bell's idea was to focus sunlight between buildings with a mirror, then talk into a mechanism through which the mirror vibrated. At the receiving end, a detector picked up the vibrating beam signal, then decoded it back into voice, similar to how a voice signal is decoded using electricity. "I have heard a ray of sun laugh and cough and sing," Bell noted in 1880. "I have been able to hear a shadow, and I have even perceived by ear the passing of a cloud across the sun's disk." Thus Bell found that clouds interrupt optical transmissions, suggesting the importance of a conduit to carry the light.

Bell's prophetic words were incomplete in one sense in defining a fiber optic system: The light source could not come from the sun; it had to be more controllable. Human beings were going to have to develop the conduit that Bell referred to, and then decode these signals.

The first big break in developing these components came some 80 years after Bell's observations with the invention of the laser, which could send signals over an optical fiber. The laser then produced the photons necessary to send information. This information flow came in the form of zeros and ones: When the light switched on, it was a one; when it was supposed to switch on but did not, it was a

[1]This section is for those unfamiliar with the basic elements of a fiber-optic system.

zero. From these patterns messages could be sent to then be discerned as voice, data, or video. "As the laser appeared, I and my associates at Bell Laboratories saw that there was in the laser as a coherent source of lightwave energy the possibility of transmitting tremendous amounts of information and that it was a valuable transmission medium," said Stewart Miller, who helped pioneer the technology at Bell Laboratories.

Interestingly photodetectors that could interpret or read beams emanating from the other end had already been discovered. "In the 1950s, there were semiconductor detectors, there was just nothing much to do with them," Miller recalled. "They were used in research work in physics—as detectors for photons."

With two of the three basic elements of a fiber optic system defined—the laser and the receiver—concentrated effort went into creating the critical third stage, the transmission medium, the optical fiber itself. The consequences of what could happen were suggested in 1966 in what became the most famous paper ever presented in the history of fiber optics. It was by Charles Kao and G. A. Hockham, both then of Standard Telecommunications Laboratories in the United Kingdom. The paper suggested that thin pieces of glass could be used as the transmission medium, and it noted that if such glass transmitted light with a loss of 20 dB/km or lower, then fiber optics would be a competitive, real-world telecommunications transmission medium.

The Kao–Hockham challenge was to launch an international research effort to make glass fiber to their specifications. A Corning team led by Don Keck, Robert Maurer, and Peter Schultz set out to find the Holy Grail of telecommunications. Their story is one of nonstop effort, sometimes using equipment that was not up to measuring the sophisticated data they were after. Through innovation, creativity, and perseverance, they were able to reach the magical 20-dB/km figure by 1970.

These elements—the transmitter, optical fiber, and receiver—were then further refined, but they constitute the basis for all fiber optic systems (see Figure I.2).

How are fiber optics integrated into an actual telecommunications line? In telephone conversations, the caller's voice is transmitted through an electronic switching circuit to program a light source to send the voice message, thereby taking the message from an electronic medium to an optical medium, that is, from electrons to photons. The associated hardware that makes this transition possible is known as optoelectronics. The optical, or photogenic, signal is carried through the transmission medium, the optical fiber, until it reaches the receiver. The latter then decodes the signal electronically and amplifies it for the listener.

An optical fiber is able to carry hundreds of thousands of such calls at one time, unlike traditional cables, which were far more limited. The optical fiber itself is housed in a cladding to contain the signal and help it travel through the core of the fiber successfully. The smaller the fiber core, the greater its ability to transmit signals distances if the core is large enough to handle the signal pulse. This was a

Figure I.2. Lasers and detectors, the electronics that send and receive signals, are packed onto printed circuit boards, then placed on racks. Here a Nortel Networks official prepares to insert a rack into an operating system. (Photo courtesy of Nortel Networks)

key advantage of single-mode fiber, whose core was substantially smaller than the multimode fiber that engineers had used.

Although early researchers faced problems funneling light from lasers to these tiny cores and attempting to splice small-core fibers together adequately, single-mode fiber became the predominant fiber type because of its ability to carry signals longer distances. Multimode fiber still has a role in today's networks, but it is generally used to carry signals far shorter distances.

The fiber optics portion of a telecommunications system has become known as the optical layer, and engineers have been trying to expand its presence to play a larger role in the network. A major obstacle to that progress is switching, which is still mainly done electronically.

Researchers realize that billions of dollars will go to those who successfully create the first optical switches. Concentrated efforts have been undertaken by major fiber optic vendors, and a number of start-ups have been created for that purpose.

Two events have substantially increased the operating speed of a fiber optic system: For one engineers have been able to develop lasers that pulse at faster rates; this is known as time division multiplexing, or TDM. Lasers, which initially

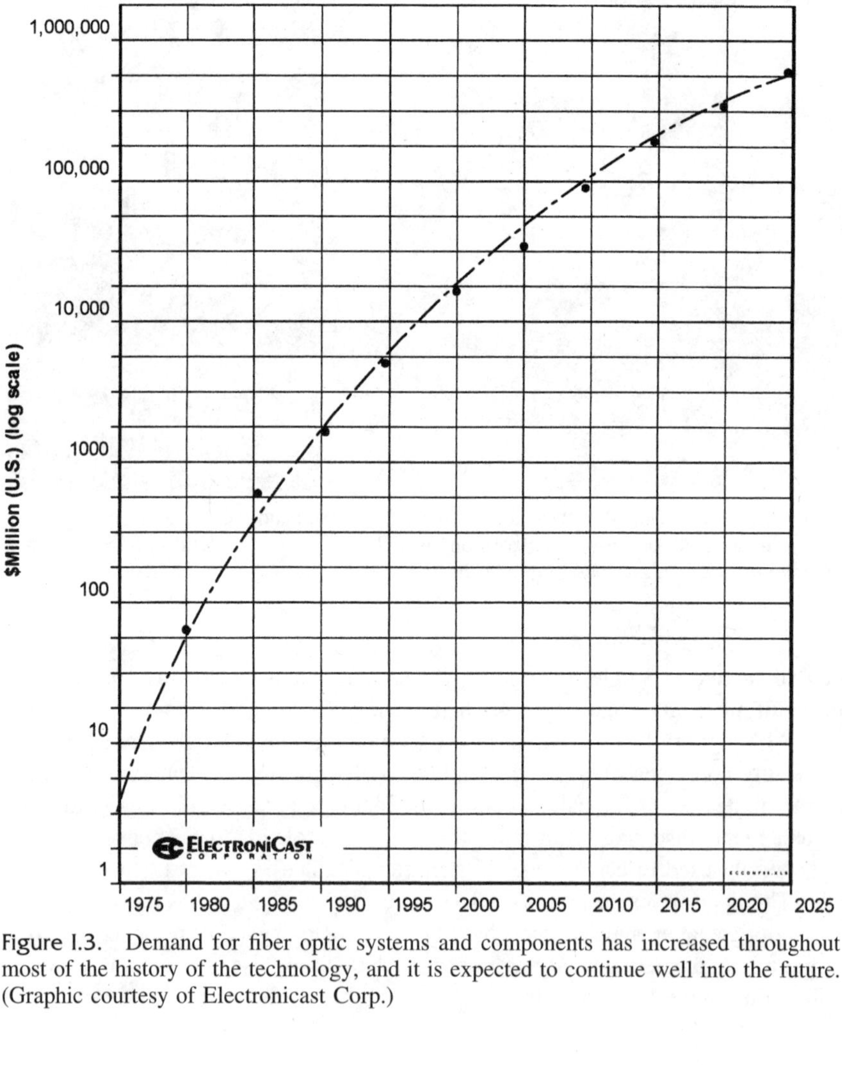

Figure I.3. Demand for fiber optic systems and components has increased throughout most of the history of the technology, and it is expected to continue well into the future. (Graphic courtesy of Electronicast Corp.)

pulsed at rates of 45 million pulses or bits per second (Mbps), increased to 90 Mbps, expanded to 270 Mbps, then to 400 Mbps, 625 Mbps, and 2.4 gigabits per second (Gbps). Currently pulse rates have reached 9.6 Gbps and 40 Gbps systems are starting to come into the network. Secondly scientists expanded the number of wavelengths—that is, information streams—carried through an optical fiber by using a process known as wavelength division multiplexing, or WDM.

While scientists have known about the potential for WDM for many years, it was not until the advent of the Internet—and the increased traffic that it brought to telecommunications networks—that WDM really came into play. At first vendors promised two channels or information streams over the same fiber. That soon expanded to four channels, then eight, sixteen, and thirty-two. Since each channel can handle a separate data stream, this in effect doubles the capability of traditional single-scream systems.

Another area of photonics growth has been photonics amplifiers—equipment used to boost an optical signal along its transmission path. The optical amplifier has to a large extent replaced more traditional regenerative multiplexers through which a signal is reformatted electronically before being sent through the optical network. The most common optical amplifiers are erbium-doped fiber amplifiers, or EDFAs, which can allow telecom systems to remain photonic for hundreds and thousands of kilometers without requiring electrons. Such optical amplifiers (also known as opamps) are critical to WDM because they allow multiple channels to be boosted without having to return to the electronic regime. Note: A fiber optic system becomes unduly complicated when every signal must be converted into an electronic domain, then reconverted into an optical domain. This is particularly true when multiple streams of 8, 16, or even more channels are involved.

Fiber optic systems have not yet failed to deliver what networks require in terms of capacity growth. A primary reason for their success is due to the amazing capabilities of optical fibers themselves, which can theoretically send multiple trillions of bits of information as needed. Demand for fiber optic systems, which will become faster and boast even more channels, is expected to increase in the future. Lasers will someday transmit 40 billion bits per second and beyond, the number of WDM channels will expand to the hundreds, and fiber is expected to continue to be more resilient and easier to use (see Figure I.3).

We should remember that the basic fiber optic system is still in its relatively early stages of development. Many new ways of expanding the capabilities of this wondrous medium will no doubt be discovered and implemented along the way.

Section 1: The Oceans As Superhighways

We do not generally think of the oceans of the world as avenues of communication. With the birth of Telstar, which used the open space of the heavens to send signals, we viewed the oceans as obstacles to be traversed, necessary encumbrances to be avoided if at all possible. This was generally reenforced by images of satellites, which work by sending signals through space to earth stations located in different parts of the world.

The problem with satellites was and is that while they may still have a role in transmitting video signals that can be used on television (a primary role of Telstar), they leave much to be desired in the field of personal communications. Anyone who has made a transoceanic call using satellite communications and suffered from mind-numbing echo—hearing the echo of your own voice—knows this problem. If we include the need for larger amounts of information, such as fax and E-mail, we encounter busy signals and other roadblocks using satellite due to its limited bandwidth capability. There is also the real possibility of a satellite malfunctioning and millions of pagers—which may carry such vital information as a request for medical assistance—simultaneously being rendered useless.

Despite the inconvenience of having to lay cable end to end on the ocean floor, cable was used as a communications form underwater for over a century. It has generally proved to be sturdy; in fact some coaxial cable placed in the 1930s is still operational between the United States and the United Kingdom, and it is used by our federal research agencies.

While it may seem easier to send messages by satellite, undersea cable has proven to be more accessible and reliable—and to provide the bandwidth required to meet today's growing information transmission requirements.

However like satellite, coaxial cable, used before the advent of fiber optic systems, also had limitations. For one thing costly and at times problematic repeaters had to be spaced closely together to boost signals continually. In contrast fiber optic systems required fewer repeaters separated by greater distances. This reduced cost, improved signal quality, and perhaps most important in an undersea system, reduced the prospect of malfunctioning equipment that could shut down the entire system and require extensive downtimes to locate the outage, dredge it from the ocean floor, then replace it.

As we mentioned, the amount of data that could be carried, known as bandwidth, was substantially greater over optical fiber cable than other transmission forms. Carriers operating submarine networks began talking about sending gigabits—or billions of bits—of information and even many gigabits of information.

The idea of carrying greater and greater amounts of data was appealing to telecommunications engineers even in the 1980s before the advent of Internet traffic. For one thing high-bandwidth networks could carry live video over cable, thus once again challenging a satellite stronghold. With cable carrying a greater amount of bandwidth, it would no longer be necessary to build so many cables as before, or so it was thought before the data explosion that increased the telecommunications line use substantially and led to the principle that new bandwidth is good bandwidth.

Before we go any farther, let us mention the effort required to test a new technology like submarine fiber optic cable.

First of all, the technology obviously had to work on land. The first terrestrial supertrunk in the United States, installed in 1981–82, was known as the Northeast Corridor project. It connected Cambridge, Massachusetts with Moseley, Virginia. As part of a concentrated national effort, the Japanese completed a nationwide fiber optic build by 1983. Three U.S. carriers completed coast-to-coast fiber wiring by the mid-1980s, as chronicled in *Rewiring*.

Yet the first transoceanic fiber optic network, TAT-8, was not deployed until 1988 due to the myriad of refinements—many undertaken by AT&T and Simplex Wire and Cable—required to place a new communications form under the sea and ensure that it would work for the next 25 years or so. Such refinements included protecting the glass fiber from constant water pressure (needless to say, this cable is far thicker than its terrestrial equivalent), ensuring sufficient power in the fiber to transmit the signal substantial distances without repeatering; and fiber reliable enough to avoid repairs.

Not surprisingly a number of smaller submarine systems were tested prior to the first transoceanic system, but the work TAT-8 did and its successful operation led to increased submarine telecommunications networking.

Rather than reducing the number of systems built, as some planners assumed, accelerating capacity requirements led to the increased use of fiber optic communications systems throughout the world and corresponding ocean fiber required to service them.

The technology continues to improve: From TAT-8's initial capabilities, follow-on systems provided far more bandwidth, increasing from 2.5 Gbps to 160 Gbps in a decade. Although on an almost project-by-project basis, naysayers criticized undersea fiber optic networks as these were being constructed, predicting that the network would not reach capacity or would never be cost-effective, every fiber optic submarine transoceanic system built from 1988–98 was cost-effective. Build the capacity and people will fill it has been the operating principle, much to the delight of those who invested in these systems.

We emphasize transoceanic fiber routes because these truly set the stage for the global information superhighway. Without such a basis, the world's telecommunications marketplace would not be emerging so rapidly as it is. It is no wonder that nations with a sea exposure are viewed as having an advantage and those without often improve their links with seashore partners to gain access to powerful international lines running to and from their continents.

The next time you make a telephone call to Ireland or send an E-mail to Buenos Aires, consider for a moment the thousands of kilometers of optical fiber cable buried beneath the world's oceans that will allow you to undertake even more sophisticated future transmissions. These oceans have become our global superhighways that make a phone call or E-mail to Japan or Portugal as close as a phone call around the block.

1

A Global Footprint

Fiber optics is uniting continents, and therefore the world, with the communications form that is making the global village once hypothesized by Marshall McLuhan into a reality.

But while McLuhan's vision was largely centered on the news business and the transmission of events that we could all share, fiber optics is providing a layer beyond that, offering individuals and companies the ability to talk to, send data to, and see each other in real time.

In reality we are crossing the oceans with a few clicks or numbers to communicate with almost whomever we want whenever we want.

Submarine optical fiber cable, now being installed every day on the ocean floor, is providing the bandwidth needed to unite the nations of the world, helping to bring down the walls that once divided people.

The day is coming when every event will potentially have a global audience, when no distance will be too far for people to span. That is due—and will be due—in large part to optical fiber pipes already in place or expected to crisscross the oceans of the world.

In the decade since *Rewiring* was published, there has been a global explosion of undersea fiber optic systems. The year that book was published, 1988, was the same year the first commercial transoceanic fiber optic system, TAT-8, was installed and operational. That was the beginning of an explosion.

In fact it led to the growth of an entirely new industry, dominated by such vendors as Alcatel and Tyco International. It also led to the construction of new equipment and ships to supply this industry.

There are now dozens of submarine fiber optic systems embedded on ocean floors. When fiber optics pioneer Charles Kao told me in 1983—5 years before the first transoceanic fiber link became serviceable—that the world's seas would some day be littered with fiber optics, he was correct. The only startling thing about his prediction in retrospect was how rapidly it has come to pass (see Figure 1.1).

The cumulative growth of undersea optical fiber cable has been unrelenting

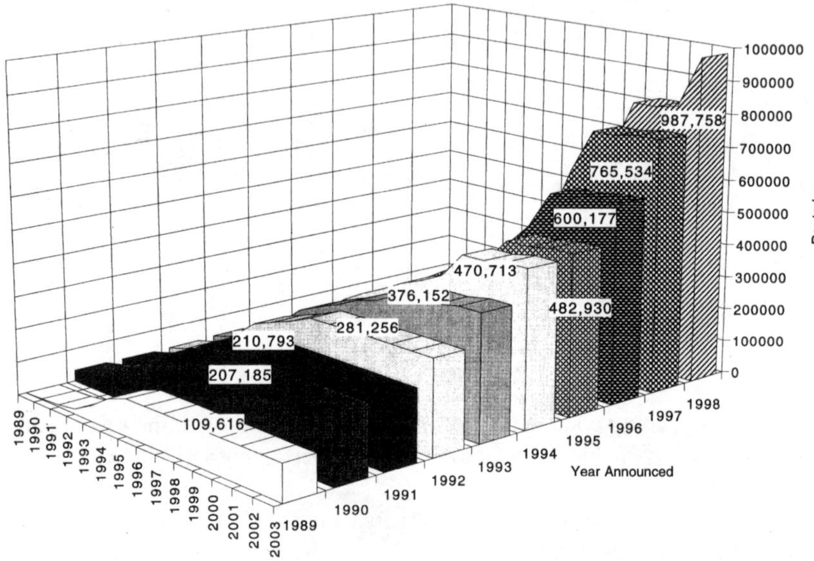

Figure 1.1. Cumulative route-km of fiber optic undersea cable installed and announced, 1989–98. (© 1998 KMI Corp., Newport, P.I., www.kmicorp.com. Reprinted with permission from the proceedings of KMI: 1998 Fiberoptic Undersea Systems Symposium)

in the 10 years since TAT-8 and PTAT-1 became operational. Why did our oceans become full of optical fiber cable? The answer is that people, organizations, and governments crave bandwidth, the ability to communicate through a variety of tools (telephone, Internet, multimedia, videoconference, fax, E-mail, and so forth), and nothing can deliver bandwidth like fiber optics.

When TAT-8 became operational, the dawn of a new telecommunications era began. Large pipes could now be spread throughout the world to allow people to communicate with each other. This change was complete: There was no crossover period when some coaxial transoceanic systems went in and slack was taken up by fiber optic systems. After the first transoceanic fiber optic system was switched on, all new builds thereafter were fiber. (The TAT-8 stands for the eighth trans-Atlantic submarine system that AT&T installed and rendered operational. The first seven were coaxial but from the eighth on were fiber optic.)

The TAT-8 was a product of the traditional telephone entities, built under the guidance of AT&T, then signing on other U.S. carriers, such as MCI and Sprint, as well as the traditional European PTTs. AT&T Submarine Systems, Inc. (more commonly known as SSI), which for awhile survived divestiture to remain part of AT&T, was the construction arm, while Simplex Wire and Cable, which like SSI was later purchased by Tyco International, handled the undersea cabling.

The traditional powers also built the first fiber optic network trans-Pacific

Table 1.1. Early Fiber Optic Transoceanic Networks

Year operational	Name	Route	Principals
1988	TAT-8	U.S.–U.K.	AT&T, PTTs
1988	PTAT-1	U.S.–U.K.	Tel-Optik, C&W
1989	HAW-4/TPC-3	U.S.–Japan	AT&T, KDD
1992	TAT-9	U.S.–Europe	AT&T, others
1992	TAT-10	U.S.–Europe	AT&T, others
1993	TAT-11	U.S.–Europe	AT&T, others

Source: Chaffee Fiber Optics

cable: TPC-5 (the fifth trans-Pacific cable and the first fiber network after four copper-based networks) was built by AT&T and Japan's international carrier, KDD (see Table 1.1).

Like earlier metallic cable submarine builds these first routes were constructed and controlled by traditional telecommunications powers to serve their primary transoceanic requirements first. Thus the globally developed nations, the United States, Canada, Japan, and Europe, were the first to reap the benefits of the fiber optic superpipes.

An enterprise known as PTAT-1 was organized by Tel-Optik, a company established by Konnie Schaefer, a submarine fiber pioneer, and two stateside partners, and Cable & Wireless in the United Kingdom. The goal was to build a private trans-Atlantic fiber optic system to compete outside the established public network.

The struggle for PTAT-1 approval in the United States lasted for almost a year and even led to hearings on Capitol Hill: "AT&T was the most vociferous opponent," Schaefer recalls, although the opposition group also included representatives from the traditional European PTTs. "Everybody filed against us; it was the world against us," according to Schaefer; "they said it couldn't be done, and we said, 'yes, it could.' "

For awhile it was "a chicken and egg thing," as Schaefer remembers. "They told us that until we had the cable-landing license we didn't have credibility, and we didn't have credibility until we had the cable-landing license." Ironically after the license was granted and the network built, Schaefer says that AT&T became PTAT's largest customer. The PTAT-1 continues to transmit traffic, as does TAT-8.

An attractive concept for PTAT-1's participants, introduced in large part by Schaefer, was the idea of condominium ownership; that is, rather than paying for the privilege of using the submarine network, the carriers actually owned a portion of it. The condominium approach helped fuel the success of such early private-line carriers as PTAT-1, and it led to an entire second tier of fiber optic submarine builds that offered alternative submarine networking to those built by AT&T and KDD. Schaefer says:

> The small carriers now had a say in the management of the pricing of capacity.... In terms of rights and responsibilities, they are very similar to the big guys.

According to Schaefer that philosophy became so pervasive that 95–99% of all undersea fiber optic networks adopted the PTAT condominium model.

The TAT-9 followed TAT-8 3 years later. (An originally planned second PTAT system never materialized.) The TAT-9 featured undersea branches on both sides of the Atlantic, connecting two points on the North America side and three on the European side.

Improvements continued as the TATs and TPCs were constructed. As a medium fiber optics continued (and continues) to improve. In addition engineering advanced, the amount of bandwidth increased, and the number of carriers also accelerated. Undersea fiber optic cable began to convince the skeptics that it was here to stay and its capabilities were growing (see Table 1.2).

With the successful operation of transoceanic submarine optical fiber cable, builders became more ambitious about size. For one thing networks became more globally inclusive, with some going beyond the single-ocean hops of TAT-8 and PTAT-1. A substantial step in the direction of global fiber optic builds, and away from single-ocean builds, was SEA-ME-WE 3, a 21,000 km fiber optic route connecting South East Asia (SEA), the Middle East (ME), and Western Europe (WE). This cable connecting Germany with Singapore features 25 landing points and 15 spurs or branches. The network originally operated at 2.5 Gbps but has the potential to carry traffic at rates of up to 40 Gbps by using WDM, which allows carriers to send multiple signals over a single fiber.

Seventy-eight carriers originally subscribed to the SEA-WE-ME 3 cable for a total price of $1.2 billion. In all, 32 landing points were planned, connecting the United Kingdom, France, Portugal, Morocco, Italy, Greece, Turkey, Cyprus, Egypt, Saudi Arabia, Djibouti, Oman, the United Arab Emirates, Pakistan, India, Sri Lanka, Thailand, Indonesia, Singapore, Malaysia, Myanmar, Brunei, Vietnam, the Philippines, Macau, Hong Kong, Guangdong, Shanghai, Toucheng, Fengshan, Korea, and Japan. The total number of people served is potentially 3 billion. Deutsche Telekom, the largest investor of the 24 carriers financing SEA-WE-ME 3, spending $63 million, which resulted in a 5% ownership share. Australia's Telstra also has one of the largest stakes, and it used SEA-ME-WE 3 to carry traffic for the 2000 Summer Olympics in Sydney.

Not surprisingly a network the size of SEA-WE-ME 3 forced controlling carriers to go beyond the one or two contractors who normally build and supply single-ocean builds. The total contractors' bill was $737 million. Alcatel Submarine Networks initially supplied 46% of the network, or $342 million, including all of the synchronous digital hierarchy (SDH) and network management for the system; KDD Submarine Cable Systems 38%, or $280 million; AT&T Submarine Systems, Inc., 13%, or $92 million; and Pirelli 3%, or $23 million.

Table 1.2. Multiple Ocean Fiber Optics Networks

Name	Length	Sponsors	Location
SEA-ME-WE 3	21,000 km	24 carriers, largest investor is Deutsche Telekom	U.K.–Egypt–India–Korea–Japan
FLAG (fiberoptic link around the globe)	16,800 km (and planned FLAG Atlantic and FLAG Pacific routes)	NYNEX, managing sponsor, others include Dallah Al-Baraka; Marubeni; Gulf Associates; Telecom Asia and GE Capital Services (now publicly owned)	U.S.–London–Tokyo
Global Crossing	Four regional submarine networks and two terrestrial	Publicly owned company	Atlantic crossing, 14,000 km; Pan-European crossing (terrestrial); Pan-American crossing, 7,000 km; mid-Atlantic Crossing; Pacific Crossing, 21,000 km; North Atlantic crossing (terrestrial–Frontier acquisition); planned East Asia Crossing; South American Crossing

Source: Chaffee Fiber Optics

Before proceeding we ask the appropriate question: What is driving the need for new global-networking capabilities? A primary answer is expansion of Internet and related data services: The growth in Internet domain hosts alone is profound and unrelenting as shown in the accompanying figure.

The Fiberoptic Link around the Globe (FLAG), which to some extent competes with SEA-WE-ME 3, connects London and Tokyo and includes landing points in Korea, China, Hong Kong, Malaysia, Thailand, India, the United Arab Emirates, Egypt, Italy, and Spain. The network, originally 16,800 km long, recently added a route between the United Kingdom and the United States.

Regional Bell Operating Company NYNEX, the managing sponsor for FLAG, is responsible for construction, sales, and marketing. NYNEX has since merged with Bell Atlantic and GTE to form Verizon. Other sponsors and shareholders include the Dallah Al-Baraka Group of Jeddah, Saudi Arabia, a large investment company; Marubeni Corp., a leading general trading company in Japan; Gulf Associates, Inc., a New York-based concern focusing on trade and project development; Telecom Holding Co., a subsidiary of TelecomAsia in

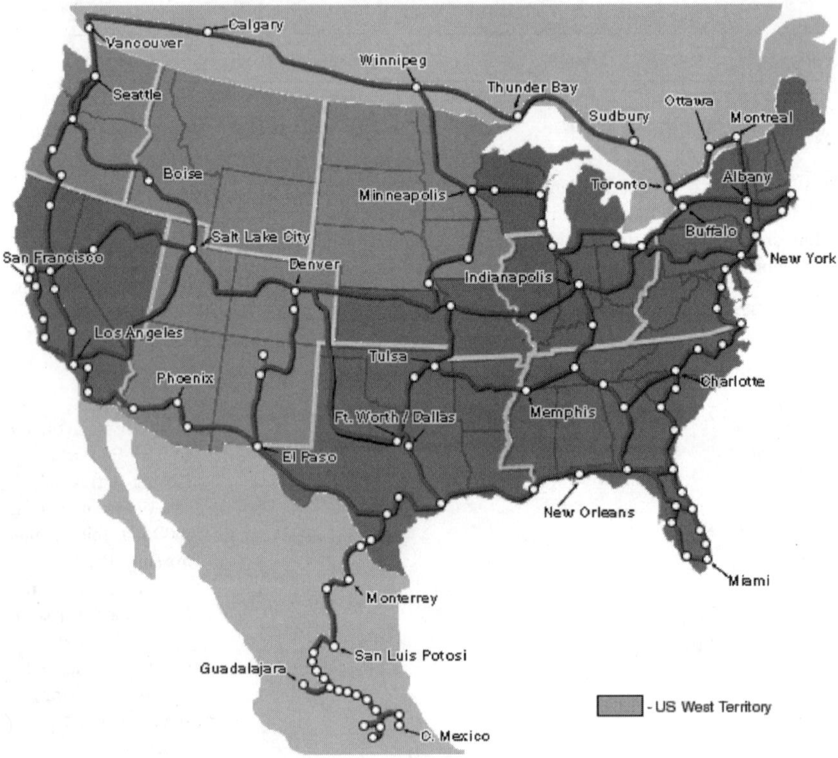

Figure 1.2. Qwest Communications has built a powerful national fiber optic network in the United States that stretches into Canada and Mexico. (Map courtesy of Qwest)

Thailand; the Asian Infrastructure Fund of Hong Kong; and GE Capital Services. Such global fiber optic builds have been followed in progressive order by even more globally oriented builds, such as Global Crossing; WorldCom's effort to build its own global network; and Oxygen, Ltd., which had not begun construction at the time of this writing.

While these networks were being constructed, a shake-up on the supplier side of the business occurred: The sale of Submarine Systems, Inc., from AT&T to Tyco International. At a time when the United States had one of the two vendor powerhouses on the scene, it was decided that AT&T would sell the venerable SSI unit, breaking up a type of monopoly but also weakening the main U.S. submarine vendor. We explore this topic in Chapter 2.

One consequence of the sale was that Bill Carter, the head of SSI, was no longer in a leadership position. As a result Carter went on to help found Global Crossing, the first submarine fiber optics venture to go public that solely concentrated on building these systems. Global Crossing consists of four regional fiber optic networks, which are at various stages of development. The most progressive is its trans-Atlantic network, Atlantic Crossing (AC-1), which came on line 26 May 1998. The full self-healing ring is now operational.

As has become the norm in the undersea business, AC-1 operates in a ring or circle design, which enables the carrier to back up traffic in case the cable goes down in one direction. This important self-healing aspect of submarine design allows a network to be self-provisioned and therefore less dependent on other submarine networks or the satellite industry to provide backup in case of system failure.

Atlantic Crossing is 14,000 km, connecting the United States with the United Kingdom, Germany, and the Netherlands. It uses WDM equipment, and it was initially provisioned to provide 40-Gbps capacity. It is upgradeable to 80 Gbps. In the design two fiber pairs carry capacity, and the other two fiber pairs carry protection capacity. For example, if traffic must be rerouted, the spare fibers can do it immediately and at full capacity.

After the initial plan was announced, Global Crossing decided to add a terrestrial link, which became known as Pan-European Crossing. To add this link the carrier gained right-of-way on VersaTel Telecom Europe B.V., providing access to Amsterdam, Brussels, and the French border. Under the agreement, VersaTel also receives capacity and dark fiber on Global Crossing's Pan-European Crossing network. With this agreement, Global Crossing reasoned, European customers could connect Europe to the United States, Latin America, and Asia. As part of the deal, VersaTel intends to build, partner, or buy around 5000 km of fiber in the next year.

In Germany Global Crossing will use GasLINE, founded by 15 natural gas distribution companies, to offer fiber optic lines and related services to telecommunications companies through existing pipeline shareholder rights-of-way, which comprise approximately 30,000 km.

The full Pan-European crossing, which continues under construction, will provide access to nearly 300,000 businesses in the Benelux, directly connect major business parks, and be connected to Germany, France, and the United Kingdom. The total business region consists of over 900,000 businesses.

Global Crossing also began constructing a Pan-American Crossing (PAC-1), a 7000-km system connecting California, Mexico, and Panama. An agreement signed on 28 May 1998, for SSI to begin constructing the network, will create the first direct path of connectivity to the United States and Asian markets for Latin American nations, says Global Crossing. Before the route's completion, Latin nations must cross the United States through terrestrial communications networks and pay a transit fee to U.S. carriers to connect to optical fiber cables reaching the Pacific Rim.

The PAC-1 system will connect California, Mexico, Panama, and St. Croix. Full commercial service is expected by February 2000. The network is initially designed to operate at 20 Gbps, but it can be upgraded to 40 Gbps using WDM. Global Crossing calls it "the largest, most powerful system to link North and Central America."

Two other Global Crossing routes include Mid-Atlantic Crossing, serving New York, Florida, Bermuda, the Caribbean, and Florida; and Pacific Crossing (PC-1), which in and of itself is a massive 21,000-km network, co-owned by the Japanese trading company Marubeni. The PC-1 was bolstered by the fact that DDI Corp., a Japanese carrier, announced that it would be PC-1's first customer. The PC-1 also uses the self-healing ring design and WDM. There are two landing points in both the United States (Washington and California) and in Japan. Initial service began in February 2000, with full ring capabilities by July 2000. The PC-1 system is designed to operate initially at 80 Gbps, but it is upgradeable to a minimum of 160 Gbps using WDM. The PC-1's organizers say it is the first noncarrier, privately owned and operated undersea cable network to cross the Pacific Ocean.

Similar in some respects to its European undertakings, Global Crossing is teaming with Marubeni in Japan to build a terrestrial fiber optic network connecting the major cities in Japan. This new venture is known as Global Access Limited (GAL), which will be owned by Global Crossing (49%) and Marubeni (51%).

The fiber optics network of some 1200 km will link Tokyo, Osaka, and Nagoya with the cable stations of PC-1, thus linking these cities to the United States, Europe, and Latin America. The three Japanese cities constitute more than 90% of Japan's international telecommunications traffic. Construction began in September 1998 on the $110 million project.

The GAL will use the latest SDH and dense wavelength division multiplexing (DWDM) technologies from leading suppliers to provide ultrahigh capacity. The network will be composed of multiple network rings for diverse routing options and restoration or redundancies among all points of presence on the network.

WorldCom originally attempted to build its own global fiber optics network, but it later decided to combine its efforts with other partners. WorldCom's trans-Atlantic system, Gemini, now has as a partner global powerhouse Cable & Wireless PLC. WorldCom's plans to build a trans-Pacific network were abandoned when it decided to join the China–US consortium, a $1.4 billion, 27,000-km network planned to connect the United States directly with China; each nation will support two landing points.

China–U.S. network planners include China Telecom, AT&T, NTT, KDD, MCI, Sprint, SBC, Korea Tel, and Hong Kong Tel. Each of the partners is expected to invest up to $100 million. The combined mission of all of the undersea fiber optic networks is to fully connect the world's communications centers to some day ensure that every citizen of this planet will have the capability to communicate with every other person. They have become the global thoroughfares, if you will, of a new way of communicating. The story behind the story, as we will see in Chapter 3, is how intelligent investors and builders have reaped the benefits of this global business.

2

The Business of Ocean Fiber

If shares were readily offered on the open market, it would have been hard to beat submarine optical fiber as the investment for the late 1980s and most of the 1990s. Every one of the submarine fiber cables that was built has costed out and will continue to provide business for many years to come, according to Konnie Schaefer, president of SwissCom N.A. (For whatever reasons a few proposed networks were never built, but those pursuing them generally lost only a small initial investment.)

Unfortunately for investors the substantial profits generated by undersea fiber optic systems were generally subsumed by larger vendors and carriers.

AT&T, Alcatel, and KDD had far larger revenue bases than their capital outlays in fiber optic systems, although undoubtedly their bottom lines were substantially improved by these activities. As with any large business, however, not all corporate activities were successful, and profits made by these systems often went to balance less successful corporate enterprises within these conglomerates.

What made and continues to make construction and operation of submarine fiber optic systems such a lucrative enterprise? The primary reason is and was the tremendous need that carriers have to provide more bandwidth: The world was becoming a smaller place, so companies had to globalize to survive. On top of this was the growth of the Internet and other bandwidth-consuming services. Companies found that additional bandwidth often allowed them to compete more effectively.

In addition to the growing demand for bandwidth, global telecom liberalization encouraged competition, thereby stimulating new markets. All this attracted new partners, new investors, and broadened market focus.

The submarine fiber optics industry was generally controlled by a few extremely successful vendors. In the early stages this included SSI, then Alcatel.

These are still the major vendors who receive hundreds of millions of dollars in orders. Alcatel was clearly in the lead when this book was written, with NEC, Fujitsu, Pirelli, and other suppliers also having presence.

The successful vendors have the actual cable-laying vessels for installing the fiber and the maritime expertise to do so effectively. A number of carriers, such as AT&T, Cable & Wireless, and KDD, also have ships. Such carriers have also logged the time necessary to ensure that their systems work underwater, a thoroughly different task from ensuring that systems work terrestrially.

Some smaller components vendors also benefitted: For years AT&T had a deal with Massachusetts-based Simplex Wire and Cable to supply submarine cable to house the fiber, which resulted in millions of dollars annually for Simplex.

Generally however AT&T (particularly before Lucent was split off) and Alcatel fed as much of the business to systems and components manufacturers within their own businesses as they could. As a result much fiber optics business was controlled by a few vendors, and such companies made out handsomely while those excluded had to rely on the terrestrial marketplace to survive.

This may have explained AT&T's reluctance to part with its SSI unit even after divesting itself of much of the remainder of its manufacturing operations.

AT&T handbuilt a successful operation in SSI that it was reluctant to lose. Under the direction of Bill Carter, SSI became a strong, cohesive workforce, known for its ability to meet challenging deadlines. Other issues surrounding the sale of SSI to another company included national security. The defense and intelligence agencies often stated that they did not want construction and sale of transoceanic telecommunications lines in foreign hands, since this would result in having to use nonsecure lines. The economic argument was that the United States should keep as much of the lucrative submarine-fiber business as possible, and in many respects national security was equated with national economic security, and that is a major battleground.

Since Alcatel was being aided by France Telecom and the French government and KDD by the Japanese government, there was some feeling that the U.S. government bore some economic responsibility to keep SSI strong.

But when TeleBermuda International (TBI) filed for a cable landing license in the United States as part of a Bermuda–U.S. fiber link known as BUS-1, it became clear that if AT&T were going to be a successful carrier, it would be at odds with a manufacturing appendage. The AT&T, looking out for its carrier interests, initially opposed the BUS-1 application in a 26 October 1995 filing with the federal government, noting that: "TBI has not demonstrated that Bermuda grants U.S. firms rights to land to operate submarine cables, as required by the Cable Landing License Act." Granting TBI an application, therefore, "would not serve in the U.S. public interest," AT&T found. However when TBI announced that it was hiring SSI to install and operate the system as part of a $45 million contract, AT&T reversed its position on 3 October 1996 with a filing made by SSI president Carter.

> AT&T SSI fully supports granting of the landing license because of our existing and potential contracts with TBI for the construction and maintenance of the cable system and because the development of BUS-1 would increase the availability of telecommunications capacity between the United States and Bermuda. (*Fiber-Optic News* [*FON*], 28 October 1996, p. 4)

This example was cited by various industry cognoscenti as proof that AT&T and SSI would be better off doing business independently: AT&T would not be shackled to a demanding vendor whose interests sometimes ran contrary to its own, and SSI would be free to seek business with AT&T competitors. "SSI did not have the freedom of action or creative pursuit of business because of restrictions from AT&T," said Schaefer.

The same issue of *FON* suggested that Tyco International, which had purchased Simplex, might be a good buyer for SSI: "If Tyco/Simplex doesn't buy SSI, they will have to compete with the rest of the world," one source, requesting anonymity, told *FON*. "They were guaranteed $15 million annually as part of the [cabling] deal [with AT&T], $15 million that may be going away." Carter himself reportedly requested a U.S. buyer (*FON*, 3 February 1997, p. 1).

AT&T did sell SSI to Tyco for $850 million. (SSI's estimated revenues for 1997 alone were in the $1 billion range. Several industry officials believed that SSI would be sold for a multiple of that, considering its specialized expertise and experience in what is an extremely lucrative market.) And there were some bumps and bruises along the way: As already mentioned, Tyco's Neil Garvey—and not Carter—was selected to run the show. The split was at first amicable, since Tyco continued to be the main contractor for Global Crossing. However, the two ended up in court against one another.

Also soon after the sale, some 140 workers were laid off as part of the Tyco–SSI merger. It is unclear how many of them were among the 800 who had come from AT&T SSI. (However according to industry officials many if not most come from AT&T.) Employees were notified of dismissal 2 weeks after the sale in what was seen as an insult to the venerable SSI and AT&T. Of course being laid off in this growth industry was not so bad as in other industries. Carter in fact became part of a new company in the field, and the industry in general was thriving.

The carrier marketplace was now much more democratic, due in large part to gains made by PTAT-1 and the rise of the condominium approach. "The whole concept was that the business approach would be different from that of the original AT&T-led consortium," Schaefer says: Carriers would buy only as much capacity as they want or need, and that would become theirs for the life of the cable. The end result was that most traditional carriers would become customers

Why have all the submarine fiber optic networks been successful? Since there was more competition, and more available bandwidth, we would have assumed numerous mergers and failures. Schaefer compares the situation to what was happening with AT&T. Ever since divestiture the carrier has been losing market share, so that by 1998, it had only 50–60% of the total long-distance business. Why then did AT&T's revenues continue to grow during that period? The answer is that the overall pie grew substantially. An AT&T official confirmed this, noting that 1997's total telecommunications volume increase over 1996 represented a larger amount than all annual traffic carried during one of the early years in the 1990s.

"The world of undersea cables has changed dramatically," says Schaefer. "We now have competition in the area of cables which did not exist 10 years ago." There still is plenty for everyone in the industry, says Schaefer, and "there is still plenty of room for others." Schaefer believes the appetite for cross-Atlantic or cross-Pacific fiber traffic is insatiable.

As a result investors from outside the telecommunication community entered this market looking for risk-free investments. Now with Global Crossing going public, people can invest in a pure submarine-fiber. That does not mean that everything was perfect for Global Crossing or that it does not have to be innovative. For example when it first went public, Global Crossing reported that the first Atlantic crossing build, AC-1, cost an estimated $750 million, but the carrier had only in the area of $400 million in binding telecommunications contracts. However AC-1 reported that another $175 million in nonbinding business, although that was still considerably short of what the network required to make a buck. Yet that did not deter Global Crossing. The carrier is already talking about building a follow-on trans-Atlantic network, AC-2, and has begun constructing Pan-American Crossing, the 7000-km cable system connecting California, Mexico, and Panama.

In fairness approximately $592 million of the $750 million total for AC-1 was met by 15 April 1998, so that it was highly unlikely that the network would not be built. "All future costs with respect to AC-1 are fully financed with the remaining availability under the existing $482 million credit facility of Atlantic Crossing Ltd.," the registration statement says. The S-1 statement does acknowledge that additional funds will be required for AC-2 or other infrastructure beyond the initial four routes.

While there is no reason to doubt the company's ability to obtain funding to construct its initial global undertaking, there is reason to question—as the S-1 form itself does—whether enough carriers will sign up for Global Crossing to make the enterprise worthwhile. The company in fact registered a loss of $10.1 million for the first 3 months of 1998; in all its net loss to shareholders reached $27.4 million. Those losses are not surprising considering the scope of the undertakings; however as Global Crossing itself acknowledges, "Success will substantially depend on sales of capacity upon [Global Crossing's] systems." There can be no assurance that any of the networks will be profitable. As mentioned earlier Global Crossing has already figured out how to be profitable: By also connecting to major cities in Europe and now Japan, in essence following its customers to major market sectors.

So many factors have changed since the first trans-Atlantic network, TAT-8, was introduced a decade ago. For one the level of security vanished as these networks were exposed to competition and deregulation and the resulting volatile pricing trends. While these networks continue to be built and to be profitable, the sure thing that AT&T and European carriers experienced with TAT-8 no longer exists. Global Crossing is betting heavily that an expanded customer base and additional traffic due to the Internet will ensure its success. Thus far it has been a winning bet, and it may remain so for many years to come. According to the registration filing:

> Much of the company's planned growth is predicated upon the growth in demand for international telecommunications capacity, which will consume the increased supply of telecommunications capacity from new cables and other technology so that price declines will not be greater than the price declines anticipated by the company in its business plan.

At first glance the list of competitors appears to be staggering. For starters there is FLAG and SEA-WE-ME 3, both of which reach beyond single-ocean aspirations. There are also the trans-Atlantic and trans-Pacific cables headed by AT&T and KDD previously mentioned, and there are Gemini, the WorldCom/Cable & Wireless trans-Atlantic venture, China–U.S., and others. Global Crossing's registration filing acknowledges that:

> The company believes that the other planned trans-Atlantic systems would compete directly with AC-1 and the commitments of the developers of these systems could substantially reduce these customers' demand for capacity on AC-1.

While things have been taking off in the Atlantic, they are expected to heat up in the Pacific. Pioneer Consulting suggests the new trans-Pacific investment will reach $4.4 billion, and it is increasing at rates exceeding 100% annually. "Existing trans-Pacific cables are unable to meet this demand even with WDM upgrades of capacity," Pioneer Consulting says.

> Capacity on the most recent transpacific cable, TPC-5, was sold out in 1997. Consequently more systems will be put into the trans-Pacific region in the next 2 years than into any other region in the world.

If that is the case, it is only because there was so much construction in the Atlantic in the past several years—with capacity still growing—as attested by Global Crossing's announcement that it intends to upgrade its AC-1 from 40 Gbps to 80 Gbps. "This new construction is being driven not only by unmet Internet demand, but also by carriers' desire to gain entry into historically under-served markets such as China," said Michael Ruddy, senior fiber optic analyst at Pioneer Consulting.

The four routes referenced include Southern Cross ($1.1 billion), China–United States ($1.1 billion), Pacific Crossing-1 ($1.2 billion) and Japan–United States ($1 billion).

As alluded to, another factor impacting bandwidth's cost—and potentially the need for these networks—is the use of DWDM, which allows carriers to use multiple streams of light along a fiber, thereby greatly expanding bandwidth capabilities. In essence significantly more data can be delivered over a fiber with DWDM than with conventional systems, which pulse only one stream of laser light. The potential risk to a company such as Global Crossing, so laden with construction costs, is that its systems may not be cost-competitive if others can provide enough capacity to render it useless.

In the end the success of Global Crossing and other multioceanic networks may very well depend on whether the information boom that accelerated such carriers as WorldCom and Qwest—and made bandwidth the telecommunications gold of the 90s—continues to deliver boundless treasure.

"While international voice traffic from 1996–2000 is expected to grow at a rate of 13% annually, international data traffic is expected to significantly outpace voice traffic growth," Global Crossing says in the statement.

> One of the key factors contributing to the growth in data traffic is the increased use of broadband applications such as the Internet, which has grown at a compound annual rate of 86 percent for the past five years as measured by the number of Internet hosts.

Additionally cable modems and XDSL are helping to fill up last-mile applications, which are expected to lead to broader demand throughout the network.

Assuming that growth continues—and this could be particularly true as Europe discovers the advantages of the Internet—and cable continues to take a large chunk of that traffic, Global Crossing could be a great success, buoyed by the demand that appears to be lifting all fiber optic undersea bandwidth suppliers to heretofore unreached heights. With so many nations beginning to demand to be included in the quest for modern communications, and with so many individuals in those countries seeing their prospects soar as they become more familiarized with these tools, global demand appears to be in no danger of slowing down.

That there are people with the vision to help them chart their future is a tribute to those people and the amazing power of fiber optics.

Section 2: North America

After the incredible surge that led to four nationwide fiber optic trunks in the mid-1980s, U.S. carriers hit a temporary wall in putting fiber optics into their networks. To be blunt this was because increases in voice-only traffic did not push fiber closer to metropolitan areas or cause more fiber to be installed throughout the countryside. Efforts on the part of carriers to try to spur customers into buying services beyond voice, such as video on demand or picture phone, did not take root.

As a result fiber optic installations leveled off and even ground to a halt. The U.S. fiber optic vendors—so quick to catch the wave of national construction—were forced to lay off workers.

One of the few exceptions to this slow down was the growth of fiber optics in the awakening cable television industry. This industry discovered that fiber made good sense in its network. Creative engineers in the late 1980s developed a way of reaping fiber's benefits by developing a new technology known as hybrid-fiber coaxial (HFC) networking. This was another of fiber's many success stories, one that literally earned the engineering crew that developed it an Emmy award.

There have been a number of important results from HFC, including the fact that elevated cable television companies are employing it in a primary position in providing advanced broadband services to the home. A more immediate result was that the cable television industry no longer had to rely on copper amplifiers to boost signals every mile or so in serial fashion. Such signals were noisy, unreliable, and depended on a row of closely spaced amplifiers. If one failed, the entire system went down. This poor networking capability represented an important reason why the cable television industry had not received high marks in its early days for network reliability or dependability and why the systems constantly seemed to be going down.

Hybrid-fiber coax networking did away with this problem by allowing operators to send signals tens of miles from the primary source without requiring repeaters. Signals were faster and quieter, bandwidth was greater, and reliability and dependability were substantially improved. The goal of each HFC network was to bring fiber to nodes connecting 500 homes, which then would be served by coaxial cable. These 500-home nodes became an important benchmark throughout the industry.

For telecommunications purposes much of Canada and the United States had already benefited from the installation of fiber optic systems in the 1980s. Saskatchewan Telephone started the ball rolling in Canada with a 3200-km fiber optic network across often rural territory. This was aided by other builds by Canadian regional carriers (known as the Canadian RBOCs), which rewired Canada with fiber optics. By forming a relationship between Canada's research arm, Bell Northern Research, and major vendor Nortel, Canadian RBOCs were often at the forefront of fiber innovation. With its national emphasis on export, and fiber optics identified as a key global technology, Canada could not afford to do otherwise.

Of the three North American nations, Mexico proceeded at the slowest pace. While some fiber optics were installed to connect northern locales with the United States and to connect some of Mexico's largest cities, the nation sadly trailed its neighbors to the north in telecommunications capacity. This gap closed somewhat when the Mexican government instituted an aggressive policy of telecommunications liberalization in the 1990s, which led to enhanced competition and laying increasing amounts of fiber.

The North America Free Trade Agreement (NAFTA) also helped provide continuity among the three nations. For example Siecor's business was exclusively in the United States in the early days, but the cabler now also includes Canada, Mexico, and Japan. "NAFTA has made for a pretty homogeneous marketplace here," says Siecor's Derek Lawrence. The design standards are pretty similar for Canada, the United States, and Mexico.

The result of opening Mexico's market was a flourish of planned nationwide fiber optic networks, including major pushes from AT&T and MCI, which grouped with Mexican partners. This liberalization also enlivened Telmex, Mexico's national carrier, which aggressively began installing fiber and joined with MCI to enter the U.S. market.

Each Siecor North American optical fiber cable order is customized, says Lawrence, although there are no extreme differences, he clarifies. For example Canada has colder temperatures, so Siecor makes materials and design choices to accommodate that. Lawrence also notes that the mechanical and environmental rigors generally are more severe in long-distance networks. The trans-Canada network for example required more rugged cable. Contrariwise when networks do not need extra features, the trend is to design as simple a product as possible to keep down costs.

A bandwidth-enhancing tidal wave hit all three nations due to the growth of the Internet and other related data services. As a result fiber optic networks that began pulsing in the mid-1980s throughout the United States and Canada grew even larger and stronger and now pulsate as never before throughout the entire North American continent. New fiber optic networks were established in all three nations as growing bandwidth demand became more pervasive.

3

The Rerewiring of America

For the most part the U.S. fiber optics industry in 1987 and 1988 was running out of steam. The four major national fiber optic constructions were completed, and suppliers were suffering through the morning-after hangover. There was no reason to push fiber closer into the cities, because there were no compelling additional services requiring bandwidth. In short the exhilarating success that the fiber industry had experienced in the United States was over, and there did not appear to be anything on the horizon to boost prospects.

The new industry had hit its first major bump in the road. Such companies as fiber optics pioneers Corning, SpecTran, and Telco Systems Fiber Optics Corporation were laying off people in substantial numbers. Boardrooms that earlier had considered how to increase fiber optic operations were now rolling back or scrapping those plans altogether. In a few short years it looked as though the industry had plateaued and was even cutting back.

In Corning's case for example it had been selling the idea of fiber optics since the early 1970s with few partakers until 1982. Then MCI requested 100,000 km, an enormous ramp-up. That was soon followed by similar orders from GTE Sprint and U.S. Telecom. However AT&T's strong resources allowed it to continue building fiber in second-tiered cities throughout the United States after its initial build was completed. A limited number of vendors profited from the first trans-oceanic fiber networks, but these were mainly captive AT&T concerns.

Competitive access providers (CAPs) were also starting to come into existence. The most startling was New York-based Teleport Communications Group (TCG), which signed important contracts with the Port Authority of New York and New Jersey, obtained right of way from Western Union, and gained funding and credibility from its association with Merrill-Lynch.

Led by former AT&T executive Robert Annunziata, TCG was a competitive access provider with staying power. (To understand subsequent events, remember that Robert Annunziata, who became TCG's CEO, had been a national accounts manager for AT&T and was able to retain excellent relations with AT&T.)

The TCG engineering was led by the ingenious Howard Bruhnke, who had 36 years of experience with New York Telephone. While a teleport usually connotes a satellite connection, TCG and Bruhnke made it clear that the basis of the venture's success was the 150 route miles of optical fiber cable Teleport had put in and around New York City. And from its origins the venture was deeply rooted in fiber, a characteristic it maintained as its operations ballooned throughout the United States. As Bruhnke used to note: "Microwave will turn your hair gray and I'm an example of that. We don't believe in microwave" (*ROA*, p. 142).

From New York City, TCG moved to dozens of cities, installing fiber along the way. The carrier eventually built a workforce of well over 1000 and served 83 metropolitan areas. And with an aggressive legal department, TCG was not afraid to compete with an entrenched carrier in the Tier 1 markets for business. In short Annunziata put together a scrappy fiber-oriented carrier with no fear of competing with the RBOCs.

In fact the CAPs, which found a variety of ways to offer alternative telecommunications were a novel U.S. telecommunications phenomenon during this period. While they had not begun to take market share from the RBOCs in the residential market, they had an impact on business customers. Unlike other CAPs, which were often funded on a shoestring, TCG was powerful and knew what it was doing from a business, legal, and technological standpoint.

Another CAP that came into existence was Metropolitan Fiber Systems (later shortened to MFS Communications). Like TCG it also believed in fiber and competing with the RBOCs; in fact it later challenged TCG as the leading CAP for primary-city markets.

Concentrating on the largest cities, these two became the leading CAPs, but dozens of others also emerged and developed over time. In fact most major U.S. cities now have at least one CAP providing alternative carriage, and some have two or more.

Perhaps no less interesting were approaches deployed by a smaller CAP, the Chicago Fiber Optics Corp., which resurrected abandoned coal tunnels underneath the windy city to reconnect office buildings with new fiber.

Other interesting CAPs and competitive local exchange carriers (CLECs) have emerged. (Note: CLEC is sometimes used interchangeably with CAP and sometimes for CAPs with their own switching capabilities.) For example Nextlink Communications, which deployed over 2000 route miles of optical fiber cable by 1998 in 32 markets. At the time of this writing the mobile telephone and satellite maven Craig McCaw was its leading shareholder. While NextLink has grown appreciably in capital investment, a satellite venture in which McCaw was involved, Iridium, is going out of business.

The GST Telecommunications, Inc., which has a major fiber optics network in the Hawaiian islands with a festooned system running among the different islands, also operates in the western United States. It claimed more than 5000 miles either built or with a lease in process by 1998.

Brooks Fiber also successfully placed itself in a number of western U.S. markets before, like MFS, being acquired by WorldCom. Intermedia Communications, which has more than 50 markets in service is another well-known CAP. The carrier is especially strong in the southeastern United States. Metromedia Fiber Network, a CLEC that made particularly strong strides in recent years, has access to valuable New York City fiber, as well as fiber in Washington, D.C., Philadelphia, and other metropolitan areas. The carrier stunned the industry by showing an early quarterly profit in 1998, and it has now decided to use its formula of concentrating on intra-city areas throughout the United States.

In the 1986–89 period, before CLECs had really geared up, before people knew about the Internet, before the Telecommunications Act of 1996 was a glimmer in Al Gore's eye, the honeymoon was clearly over for the fiber optics industry. Nationwide fiber trunks had alleviated most intercity problems, and capacity build-up had not really challenged metropolitan areas so that deploying fiber optic was worthwhile. Although seeking ways to use fiber, the cable television industry had not found a networking solution to that effect. There was no demand for more capacity: People would not pay for new fiber construction when their main, and often only, telecommunications interest was using the telephone.

Symptomatic of what was going on was a report published by Spectrum Planning, Inc., which described the radical plummet in fiber deployment befalling RBOC Pacific Telesis, US West, and BellSouth in 1986—as compared to 1985 levels:

> More than anything, we believe that the exceptional crash in the fiber requirements as set forth by these RHCs is a reflection of the coming to a close of the long-haul buildup. (*FON*, 31 March 1986, p. 1)

Yet throughout the history of the fiber optic industry, ingenious and creative minds have overcome numerous obstacles because fiber has the best characteristics of any communications medium. In fact the basic properties of the medium have been too good to interfere with its continued presence and growth. Remember that compared to most communications technologies, fiber optics was still fairly new in the late 1980s. The challenge was to develop this technology when near-term hopelessness set in.

A critical avenue opened for fiber in 1989 with the advent of the first HFC systems for cable television. Long plagued by noise and distortion traced directly to the copper repeaters and cable ubiquitous throughout cable television systems, engineers tried to introduce the many advantages of fiber into the industry to help address some of its obvious problems. For an industry that had at times been criticized for its outages and general unreliability, fiber offered the potential for secure, high-quality transmission if cable television engineers could figure out how to run cable television signals over it.

The ultimate solution perhaps could be to run fiber to every cable television residence in the United States eventually. However that was impractical at the time, since fiber was too expensive and too difficult to work with. (Note: Since cost and difficulty were viewed as two barriers to the further advance of fiber, much time was spent in the late 1980s to reduce costs and make fiber easier to work with. These challenges continued in the 1990s and will no doubt continue into the millennium.)

In the meantime however the goal was to bring the advantages of fiber to cable television networks in as cost effective a manner as possible through the advent of HFC. The HFC was developed at Time Warner Cable (TWC) in the late 1980s by a team of engineers working on analog fiber technology, that is, transmitting analog video signals over an optical fiber cable. Key to this effort was Jim Chiddix, who was still at Time Warner Cable when this book was written; Dave Pangrac, who has since formed his own consulting firm; and Louis Williamson, who was also still at Time Warner at the time of this writing.

The concept of analog fiber transmission was demonstrated in the laboratory in 1987 by Time, Inc., recalls Chiddix, before Time merged with Warner (Chiddix has since risen to the position of senior vice-president and chief technical officer for Time Warner Cable). The first commercial application was a fiber trunk in Orlando, Florida in September 1988, which used analog lasers supplied by what became Lucent Technologies. According to Chiddix:

> It performed extremely well and, in fact, in fairly short order, our people in Orlando decided they were going to use the fiber as the primary link and microwave (which had been the primary link) as a backup.... Raindrops absorb microwave radiation and you lose the signal sometimes.

Before HFC fiber optics were used to transmit digital information, lasers, turning on and off in rapid succession, and signals constructed from those pulses were used to deliver very high bit rates in long-haul applications. The breakthrough came when the laser light was turned on continuously and modulated, with the entire radio spectrum carried on coaxial cable to the home, says Chiddix.

In many respects Chiddix was already familiar with fiber's many advantages, which proved to be an important advantage. He had worked with fiber at American Television and Communications (ATC) in Honolulu, where ATC installed some fiber links on the island of Oahu. When Chiddix came to Time in 1986, he met Pangrac, who also had some fiber experience. Chiddix hired Pangrac, brought him into the corporate office, and from there they assembled a team trying to improve how an HFC system works. In Chiddix's opinion:

> It was clear that cable technology was running into some fundamental limitations which were architectural ... To go any distance at all, you have to put in a coaxial cable amp every thousand or two thousand feet; that provides a real reliability limitation.

(Note: Each of the amps also added distortion and noise.)

Pangrac also recognized the enormity of the problem: "The coaxial equipment was so bad we had thousands of people unhappy with it." He recalls that:

> The main problem was the long line of copper repeaters that had to be replaced every mile or so. Our feeling was that if we could locate fiber in the middle of that, if we could break up the long line of noisy copper repeaters, then the industry would benefit tremendously.

Although coaxial cable is capable of carrying more than a gigahertz of bandwidth, to ensure usable signal quality at the end of the coaxial system it was necessary to greatly reduce the number of channels carried, recalls Chiddix. "We were looking for a way to change the basic architecture." In fact Chiddix, Pangrac, and other team members presented a host of papers at succeeding cable and IEEE electro-optic device technical meetings that basically lit the HFC fire in the cable television engineering community.

Chiddix and Pangrac noted in one paper that:

> Optical fiber transmission technology has achieved rapidly increasing acceptance by the cable television industry.... While 87% of the homes in the United States are already passed by a broadband coaxial CATV network, coaxial technology, as it is presently being used, is beginning to approach its performance limits. Optical fiber offers a high-bandwidth, low-loss transmission medium, which holds out the potential to allow significant performance improvement in today's cable television networks.

Chiddix and the team "began prosteletyzing this technology to get the cable engineering community excited about this." They appealed to the laser community to invest in making a better device. The AT&T became interested, as did John Egan from Antec, a major distributor of AT&T. "Egan introduced us to AT&T optics people," remembers Chiddix, "he was a big cheerleader for this technology."

Some of the first lasers were made by Ortel Corp., which used them in a similar military application to provide remote fiber feeds to satellite dishes on battlefields. "We saw this as a transforming technology," says Chiddix, who remembers Herzel Laor as an important consultant who joined the team, but not with exactly the same reverence as Pangrac does: "We developed the concept, with the help of Herzel Laor, who was our teacher." From these sessions the idea of using fiber effectively in an analog radio frequency (RF) world grew to become a reality. Pangrac recalls

> Making early presentations to that effect and having important people, including an engineer at one of the leading MSOs, get up and chastise us and tell us it was impossible.

Chiddix does not believe there was open opposition to the HFC architecture, "just a lot of inertia. We gave hundreds of talks ... it was like turning a supertanker."

Chiddix also notes that "there were a whole school of people that didn't trust fiber." In one instance, he recalls, they installed a fiber network that had to include a switch to go to coax if the fiber failed. "It was like gasoline cars having to still have a holster for the buggy whip."

There was opposition to this work even at ATC, one of the places the technology was being developed, however once the technology began to work, the climate changed. "The people who threw us out earlier were now were working for us," recalls Pangrac, who also credits Larry Stark, now with Ortel, with helping to establish the importance of the technology at ATC.

Pangrac and colleagues took the concept to Scientific-Atlanta and proved it to Jim Loveless there. "We asked them to challenge it," says Pangrac. "They ran it and it worked fine." As a result Scientific-Atlanta began to produce some of the first commercial HFC systems. A new fiber medium began to be used in cable television systems throughout the United States, which improved their reliability and thereby the industry's reputation. However just as importantly, fiber provided the cable television industry with a launch pad for future services. (General Instrument also made amplifiers and a variety of equipment for the project, recalls Chiddix: "They glommed onto this.... They teamed up with Ortel, particularly on the laser technology.")

Scientific-Atlanta was a little slower into it, although the company eventually did catch AT&T, says Chiddix. "They had a handful of distributed-feedback lasers with very narrow linewidth and they didn't chirp," he says. Chirping occurs when lasers change wavelengths while transmitting data streams. It is unacceptable because the data stream can combine with another wavelength. Early lasers were not easy to make, concedes Chiddix. "What AT&T found was that if they hand-selected them, they could find enough lasers that worked. GI was using Ortel lasers and Scientific-Atlanta was using Japanese lasers."

The idea of the 500-home node came a little after the fundamentals of the technology were nailed down. "In upgrading existing cable systems, you can push fiber as deep as you want," says Chiddix, "we looked at fiber to various depths. We just wanted this all to work better."

Not surprisingly Time found that bringing fiber close to the home was "very capital intensive, and we built this business by being thrifty," says Chiddix. From his experience:

> If you run fiber to within about a kilometer to people to the end of the coax, you are just beginning to go up the knee of the cost curve. If you don't go that deep, you don't save as much. That one spot is where you get a lot of benefit, but not a big premium. In average density (relatively normal dwelling spacings), that involves plant that yields about 500 homes.

No matter where the fiber stops, there is always the potential of taking it closer to a home if the need arises. "We run six fibers to the node, even though we only use one pair of fiber.... The others are spares, or there for expansion," says Chiddix. Five-hundred-home nodes work well for cable modems, which are propelling HFC upgrades. Due to demand for very rapid Internet connections, optical fiber must send signals both upstream and downstream, unlike the more conventional one-way cable television systems. The fiber portion of the network used "plain old single mode fiber, which was very well suited to this," according to Chiddix. For fiber to be cost-effective closer to the home, the cost of the electro-optics must be reduced, Chiddix says. "On the other hand, the cost of optical fiber cable is very cheap," he adds, costing some two cents per fiber foot.

Cable television operators value the fiber that has the lowest cost and best splice performance. Siecor Senior Vice-President Derek Lawrence says that:

> In the cable TV area, they have tended to stay with single fiber splicing, using analog equipment, where in the telephony area, they have deployed exclusively digital, generally mass fusion splicing.

In recent years, the cable television market has exploded to the point where, in some cases, it is larger or as large as the telephone market, says Lawrence. In some instances cable television customers have recently been Siecor's largest customers, including the large telecommunications carriers.

Time Warner Cable intends to have 70% of its network fully HFC by the end of 1998 and the entire job completed by the year 2000. Because of what its success meant to the entire cable television industry, the Chiddix Time Warner engineering department received an Emmy award due to its role in "developing and advocating broadband fiber optic technology and architectures" (see Figure 3.1).

The growth of synchronous optical networking (known as SONET), another key networking advance first developed in the 1980s, provided important benefits for fiber optic networks in the 1990s; the international equivalent is SDH. The first development in this area was introduced in lightwave systems, which were unlike their fiber optic predecessors because these were "no longer photonic clones of their electronic predecessors," says Kathy Szelag, director of fiber optic products for Lucent Technologies. Instead these included real optical multiplexers, which began at very low speeds. The first lightwave systems were introduced around 1984.

At that time engineers discovered that fiber optic bit streams could be transmitted more efficiently through synchronous transmission rather than asynchronous, which was the form used earlier in the 1980s. Before SONET, vendors built their own systems, which often had little to do with competitor fiber optic systems. Szelag remembers that:

> The technology was changing so fast.... The first systems that went in about the 1983–84 timeframe were operating at 45 Mbps; the second group operated at 90 Mbps; the next systems pulsed at 180 Mbps.

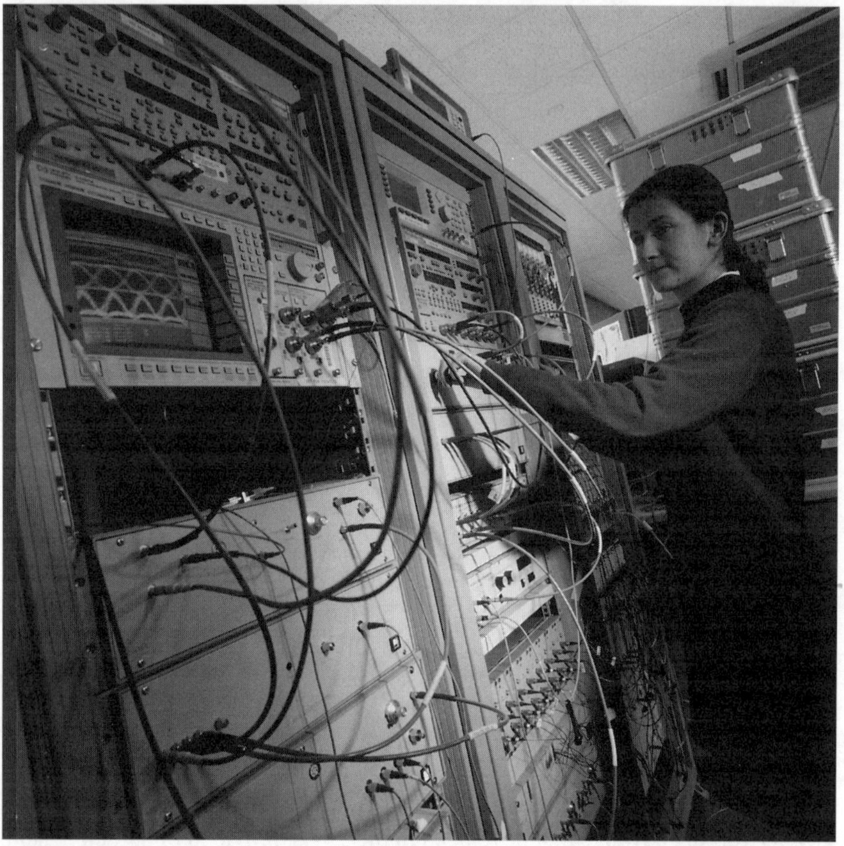

Figure 3.1. Characteristics of a laser signal being monitored on the screen to the left. The technician adjusts the signal for even greater clarity. (Photo courtesy of Nortel Networks)

After that there was a divergence: AT&T's manufacturing arm, what is now Lucent, developed systems operating at 417 Mbps, Telco Systems and Fujitsu made systems operating at 565 Mbps. "They simply didn't agree on line rate," Szelag noted.

> However, we all did agree that what we had was a technology that showed lots of promise.... This was the future. We decided to go to the standards body, assuming the line rates will go up right away, and plan a set of standards. The primary goal of standards, the primary purpose of internetworking, is to make sure that everyone is working together for the betterment of the network.

The first manifestation of this was SYNTRAN (synchronous transmission), which came into being in the mid-1980s. Its goal was to provide coherent standards to an industry that was lifting off in a hurry, recalls Szelag. The SYNTRAN represented "a dramatic change in the way these systems were built," says Szelag, "the clocking mechanism was internal as opposed to external. The primary reason was so that the lightwave system could clock itself." Unfortunately that led to disagreements among the participating vendors, so SYNTRAN as a standard died almost immediately.

From that background came SONET. Fujitsu was the first to build to its standards. However there was a split almost immediately between what the United States and Canada wanted and what other nations desired. "It was extremely complex; governments could not agree on a common standard," recalls Szelag. As a result SONET was developed for the United States, Canada, and a few other nations, while SDH became the standard for most of the world.

Szelag acknowledges that AT&T was not so aggressive in formating the first SONET standards as it could have been. "We knew the standard was so young that Phase 1 was not a good thing to be building to; it was too immature," she remembers. In fact Fujitsu, which had pushed the initial standard, had to abandon it.

During the second attempt to agree on a SONET standard vendors really came together, and the standard as we know it today began take shape. According to Szelag:

> It was a bold look into the future, but they decided to map out transmission rates not only for now but for what was to come. They began to standardize on 155 Mbps, 625 Mbps, and they also picked up 2.5 Gbps and 10 Gbps, anticipating that it would take off, which it eventually did.

The SONET rings, which are now a ubiquitous part of the standard, were not included in the original SONET standard, they were developed in 1989. "The North American market led the way in adopting synchronous optical network standards," says Greg Wortman, who spent years as senior director of marketing at Fujitsu Network Communications, Inc. The SDH deployments came about 2 years after SONET, Wortman calculates.

Bellcore (now Telcordia Technologies, Inc.) was the primary motivating force behind the SONET standards in the United States, while the CCITT was the forum through which the SDH standard was developed. Various vendors contributed features to the evolving SONET standard as guidelines for power levels, craft interfaces and SONET rings evolved. Nortel for example developed the line-switched ring. As earlier Fujitsu took the lead in a variety of areas.

An important element of this SONET evolution was digital cross-connect systems (DCS), which connect SONET equipment to the outside switched network. AT&T initially took the lead with DCS, but it did not keep up as new iterations entered the marketplace. "We made a mistake in the marketplace when we decided not to add SONET functionality to digital cross connects," acknowledges Szelag. As a result a smaller vendor, Lisle, Illinois, based Tellabs, became the DCS market leader. Tellabs has grown significantly thanks to the success of SONET and DCS and it became a major success story as a U.S. telecommunications company.

Fujitsu decided that it was not large enough to service both the long-distance and local-exchange markets, so it decided to concentrate on only the SONET local exchange market in the United States, says Wortman. The company was the first to market with SONET add–drop multiplexers operating at the OC-3, OC-12, and OC-48 levels. As the result it was no surprise that "we were very successful at capturing the Bell operating company market," Wortman says.

"It was purely a decision of the best way to grow our business," recalls Wortman, of Fujitsu's decision to focus its resources in the United States. "The decision was made that the combination of a very large steady SONET market would exist first with the LECs and that the longhaul carriers tended to be more cyclical."

Fujitsu's customers included virtually all of the Bell operating companies, embracing Bell Atlantic, NYNEX, Ameritech, Bellsouth, Southwestern Bell, and US West. Fujitsu was also involved in implementing the ring standard, which was important for the concept of survivability and a major advantage for fiber optics as a technology. If there was one downside to the amazing bandwidth an optical fiber cable could carry, it was that a lot of communications was lost when an optical fiber line went down. Backhoes in particular were a major scourge to the industry. A SONET bidirectional path switched ring (BPSR) suggested survivability because traffic could be backed up and routed the way the signal had come, and then around another way to its destination.

Fujitsu was a major success in the United States, where barriers to entry were relatively low, Wortman observes. That was not the case in Europe, where barriers were often politically based. "We as non-Europeans have not been as successful because the barriers to entry really have been political," Wortman says.

Critical to the success of SONET were Fujitsu, Bell Labs, and Bell Northern Research, which Wortman characterizes as "the three largest private telecom labs in the world." Much work was accomplished at both the high-end and low-end of the SONET market that first began in those laboratories, he adds. While some research and products destined for the U.S. market are still centered in Japan, Wortman notes that Fujitsu now employs 2400 people in North America, and it has a major manufacturing center and five development centers in the United States. The vendor has engineering centers in New York, Texas, Massachusetts, California, and North Carolina.

A more recent SONET product that such vendors as Fujitsu, Hitachi, Nortel, and Lucent were involved with is bidirectional line switched rings (BLSR), which are much more complex to employ than the unidirectional line switched rings but make sense to use when heavy traffic is not evenly distributed around the ring.

This particularly makes sense in most long-distance applications: "The pattern of traffic in long-distance SONET fiber optics lends itself to BLSR," says Wortman. This was an emerging SONET market in the late 1990s.

The SONET was a very successful standard considering the number of standards attempted. "It has accomplished what it set out to accomplish," says Szelag. It set out criteria that at times were difficult but never impossible to adhere to. She believes that SONET will eventually disappear, although its growth is substantial. In fact Szelag calculated in 1998 that the market share for SONET products increased an average of 15% annually in the 12 years of its existence. Price pressure in the industry drove down the cost of SONET and SDH equipment.

There are already future scenarios in which SONET does not occur in systems, and in fact a few carriers were experimenting with such systems as this book was being written. One inherent problem that SONET faces is the fact that it was designed primarily for voice-oriented networks, not for the growing data requirements that now prevail. "SONET delivers high-quality and highly reliable voice; however, it was not designed for data," Szelag says. Nevertheless no other standard has yet been able to provide such high transmission, restoration, and high-quality voice as the SONET standard has. On the horizon is Internet Protocol (IP), which is more focused on data. IP over photons represents an alternative to SONET that many in the photonics community believe will diminish the importance of SONET and SDH. Szelag points out that interoperability among vendors did not really begin to occur until the summer of 1997. That effort is now fully underway.

Another new photonics technology changed the way fiber optic systems worked by successfully using the very strong attributes inherent in a photonics system: Scientists from AT&T Bell Laboratories and other facilities long hypothesized that a basic optical fiber could carry more data than what one laser transmits, since these were streams of light made up of massless photons. This could be effected by sending multiple beams over the same fiber at the same time. Such a phenomenon is known as WDM, which involves sending multiple waves or colors over the same fiber to produce many different transmission streams. Thus an optical fiber that once carried one group of message signals could be transformed into a medium that carried two, four, eight or more (see Figure 3.2).

The economic impact of WDM was profound: Rather than building new optical fiber to provide more capacity, carriers could attach WDM electronics to both ends of the fiber, thereby multiplying the number of signals being transmitted—exactly the effect of adding more fibering.

The implications are also important for the fiber industry. While having to relinquish some of its early patents, and the licensing that accrued from them, simply because they are expiring, Corning is concentrating on a new area of intellectual property that deals with better transmitting multiple signals over an optical fiber. This is related to the introduction of its Leaf fiber, which has a larger

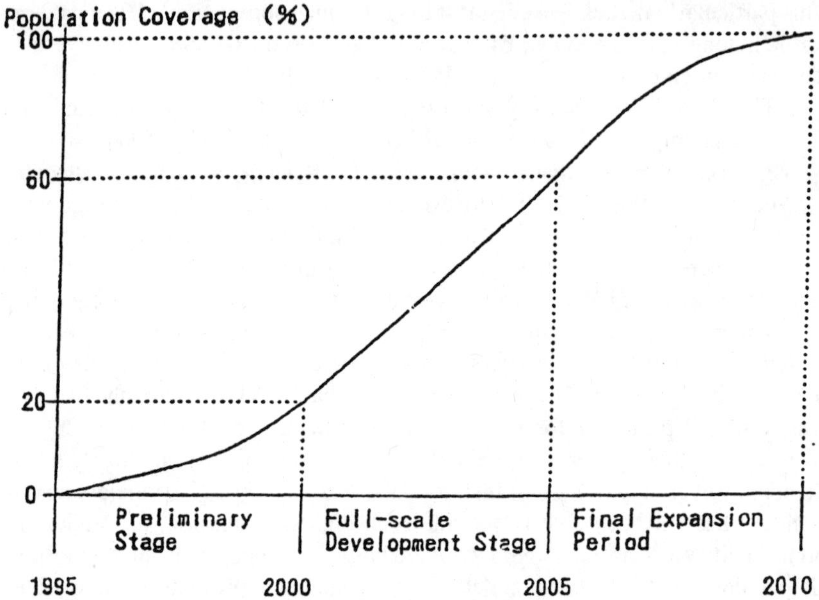

Figure 3.2. A page from an early planning document of the Japan Telecommunications Council calling for nationwide wiring with fiber optics to be completed about the year 2010. (Courtesy Telecommunications Council [Japan])

core diameter for delivering the power necessary to propel multiple wavelengths through a fiber. "We certainly have a lot of patents for higher end products geared to the WDM marketplace," Corning patent attorney Mark Laoeresch says.

Pirelli was the first vendor to offer a successful commercial product in this field. This was a 4-channel WDM system introduced in 1994. Lucent Technologies says it began delivering 3-channel units in 1995, but these were almost entirely to former partner AT&T.

The DWDM industry really took off when a new start-up, Ciena Corporation, began delivering 16-channel systems in early 1996. These were known as DWDM systems because the wavelengths were closely packed together in the spectral band.

The story of Maryland-based Ciena represents many things. It is the tale of what a small, focused company can do when it has the right technology at the right time. The tale also reveals how the fiber optics industry caught its second wind and lifted some heretofore unknown entities to unimagined heights. But it also is the story of how a small company can become vulnerable to larger, more powerful companies and how a corporation can be seduced by the financial community and the prospect of amassing individual wealth to the detriment of building a company into a successful enterprise after its terrific lift-off. Unfortunately, while Ciena had the potential to become another major U.S. telecommunications force, it struggled to maintain its early success before reaching a sound growth path.

Founded in 1992, Ciena's roots came from an idea by David Huber, whose original idea was to transmit multiple television channels over an optical fiber using WDM. Huber went to a variety of potential suitors for funding before contacting the Dallas-based venture capital firm of Sevin, Rosen. But before Sevin, Rosen would provide the required capital, the company told Huber that the fledgling enterprise would need a new CEO to run it and shape the business plan.

This was to be Patrick Nettles, who, like Huber, had a Ph.D. and had worked in the broader area of telecommunications. The affable Nettles, who had considerable executive experience at the start-up level, tailored the business plan to match WDM to the telecommunications marketplace, which he believed was growing faster and stronger than the cable television market.

No one could have predicted the serendipity that followed, but such events launch successful businesses. At about the same time that Ciena was developing WDM technology, long-distance carriers found that their fill rates were rapidly increasing. As Ciena likes to say, they were becoming "bandwidth constrained." This represented a new opportunity for Ciena's multiwavelength solution.

After a number of years of dormancy, the need for more bandwidth was being shaken awake. There were a number of driving factors, by far the most important was the need for new data-oriented services rather than voice over the network. People now used the fax machine with regularity, they were becoming familiar with personal computer on-line services. Largely through the efforts of Andy Grove's Intel, personal computers were becoming more powerful, with the ability to capture and send more information over telephone lines, and—thanks largely to Microsoft—they were using more sophisticated software to interact with these services.

For years carriers dealt with only small, incremental voice-growth patterns, now they faced major bandwidth constraints as a result of Internet-oriented bandwidth demand. In looking for tools to provide the most capacity in the shortest period of time, carriers found Ciena, which now led the market with its 16-channel DWDM systems. And Ciena enjoyed the brief golden age that all vendors dream of: The best product, highly in demand, which no competitor could duplicate.

As telecom forecaster George Gilder observed, the demand for bandwidth could far exceed the requirements of Moore's law, which predicted that the number of transistors on a microchip would double every 18 months. Gilder's hypothesis, the law of the telecosm, states that bandwidth will triple each year for the next 25 years, creating trillions of dollars in new wealth.

As Gilder correctly points out, the reason that WDM was able to take off was the advent of the EDFA. Such a device allows signals to be boosted optically rather than by the earlier, bulkier regenerators that formerly converted them to the electronic domain, and then back into the optical domain. The result is cheaper technology delivered through the far more efficient optical-networking domain, often referred to as the optical layer of the larger telecom network ("Fiber keeps its promise," Gilder Technology Report, February 1997, PL.).

In fact WDM and DWDM represented an important departure from the traditional way telecommunications managers increased capacity in a fiber optic system. The first systems began operating at 45 and 90 Mbps using one laser before increasing to 2.5 and 10 Gbps. This process of making faster lasers and detectors to increase a fiber system's capacity was known as TDM.

Another golden window of opportunity in the 1990s was Nortel's production of OC-192 systems operating at 10 Gbps. Despite objections from Lucent and other vendors that lasers operating in the Nortel system would have problems with traditional optical fibers handling these faster data rates, Nortel plowed ahead and gained a 2-year window in bringing these rapid systems to the telecommunications marketplace.

Rather than simply increasing laser speed, WDM allowed carriers to send more signals along the same fiber. While the next increment in TDM to 10 Gbps increased capacity fourfold, Ciena's 16-channel units boosted that capacity further to 40 Gbps. This was accomplished by creating what it called virtual fiber; that is, transmitting over one fiber what it had earlier taken many fibers to achieve (see Figure 3.3).

Whether Gilder's prediction comes true—he has engendered skepticism even among some of the most ardent fiber proponents—Internet and associated services breathed new life into the U.S. fiber optic industry. And when you combine that with the competition that the Telecommunications Act of 1996 introduced, you have an explosive growth situation. This situation led to yet another rewiring of America, this time with fresh fiber by an entirely new group of carriers. While Sprint and MCI once considered themselves alternative carriers to AT&T, they are now in the category of traditional carriers as such new forces as WorldCom, Qwest, IXC Communications, Williams, Level 3 and Frontier built nationwide U.S. fiber optic supertrunks in the 1990s.

The more traditional long-distance carriers have not packed up their marbles and gone home: AT&T, which won the 1980s race to be the first continental U.S. coast-to-coast fiber carrier, carried 97% of its traffic over optical fiber cable by 1998. Its entrenched base of more than 40,000 route miles of optical fiber cable is intelligently engineered into 20 national SONET rings to provide a highly protected network.

Remember Annunziata's connection to AT&T, mentioned earlier in this chapter? It was cited as one of the reasons why AT&T bought TCG in early 1998. In doing so, AT&T had one answer to the local services business it so desperately craved. Said AT&T Chairman C. Michael Armstrong:

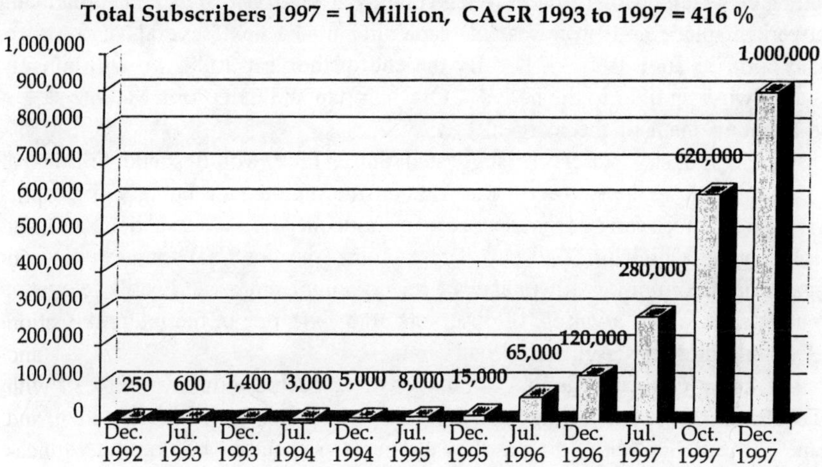

Figure 3.3. The United States and Europe are not the only places that have seen drastic increases in Internet use. The trend is expected to continue. (Analysis and representation courtesy USITO; sources of data: CNNIC, MPT, ChinaNet, and AT&T)

> TCG has more fiber route miles and serves more businesses in more cities than any other competitive local services company.... The strategic value of this merger, combined with other initiatives we are undertaking, positions AT&T for growth and undisputed leadership in three of the fastest growing segments of the communications services industry—consumer, business and wholesale networking.

At the time of its acquisition by AT&T, TCG had installed more than 10,000 route miles of optical fiber cable—reaching 508,283 fiber miles—and 50 local switches. (Note: Fiber miles are the number of route miles of fiber cable times the number of fibers contained in the cable. For example if a single-fiber cable travels 1 mile it goes 1 route mile; if that same cable contained 16 fibers, it goes 16 fiber miles).

But AT&T was far from done. In an announcement that surprised the telecommunications and cable television industries, megacarriers AT&T and TCI announced that they would join forces. The announcement represented many things, not the least of which was a victory for HFC as an architecture for bringing fiber into the home. The integrated AT&T–TCI network relied heavily on HFC: the AT&T used HFC to gain an edge on the RBOCs.

The union also joins the single largest fiber optics carrier in AT&T—although that is disputed since MCI and WorldCom joined forces—and the leading cable television provider, which claims to have 60% of its HFC broadband network in place and promises full capability in the next several years (fibertoday.com, 25 June 1998, p. 1). "By the end of the year 2000, we will be fully rebuilt, two-way fiber to the node," TCI Chairman and CEO John Malone said at the announcement of the merger.

Some industry observers suggested that AT&T would spend billions of dollars integrating the two networks. Yet considering the fact that AT&T's access to local residences can go from near zero to about one-third of the U.S. homes, that might not be a high price to pay. There is also the potential for new kinds of engineering, combining advantages of telecommunications and cable television architectures, which a variety of observers noted was one of the intentions of the Telecommunications Act of 1996 all along.

In fact a consensus among a number of respected industry sources is that AT&T and TCI are attempting something far more in the spirit of the act than what some of the RBOCs are doing as they attempt to expand by combining with one another. The AT&T–TCI merger was to receive full approval, although in some ways it has caused AT&T to itself consider splintering into several companies.

Sprint, whose pin-drop quality ads in the 1980s visualized the national movement to fiber optics by demonstrating the clarity that fiber optics defines, was the first domestic carrier to recognize the potential of Ciena Corp. Sprint worked with Ciena, allowed it to use its labs extensively, told Ciena what it required, and helped Ciena go into business with its 16-channel systems. As a result Sprint became the first U.S. carrier to use the 16-channel systems on a regular basis.

The third carrier to construct a 1980s national fiber trunk, MCI Communications, decided to join the new-generation fiber optics carrier WorldCom in an exciting union that created at least the second largest U.S. long-distance carrier. As we were going to press, Sprint was also attempting to join the party. The combined entity says it would have more than 45,000 miles of route fiber cable, enough fiber to stretch between San Francisco and Washington, D.C. 16 times. The combined carrier would have annual Internet sales of more than $2 billion—and grow at the rate of 70%.

The union of a new fiber optics such as WorldCom, which prides itself on putting in the newest networking, and MCI, which is still trying in some respects to obtain the most from its entrenched fiber base, should be interesting. One thing is certain: Both achieved their current positions by being fiber innovators, and that trend is expected to continue.

New national fiber optic carriers have enlivened the marketplace. Qwest Communications is building an 18,449 route fiber mile network, the Qwest Macro Capacity Fiber Network which will embrace 130 cities in the United States. It has the capability of offering up to 2 terabits per second. Remarkably Qwest reported a profit in the fourth quarter of 1998 and it anticipates profitability in calendar year

1999, a tremendous feat considering the capital investments required to building a fiber optic network (see Figure 1.2).

The Qwest network consists of 48 fibers buried 4–5 ft underground, primarily along secure railroad rights-of-way, the company says. It runs at speeds up to OC-192, and it has DWDM capabilities. The carrier also demonstrates marketing savvy by doing business with a number of different information-intensive groups and supplying the bandwidth they require through a variety of different scenarios ranging from carrier's carrier to services provider.

Another important construction comes from Texas-based IXC Communications, Inc. One goal of IXC was to reach 13,000 miles of active fiber route miles by the first quarter of 1999. The network also hired an impressive array of executives, led by Ben Scott, to run the project (see Figure 3.4). It has since partnered with Cincinnati Bell to become Broadwing.

Still another national construction is being undertaken by Frontier Corp., which is constructing a 15,000 mile ring-based fiber optic network. This configuration includes three western rings connecting Seattle, Portland, San Francisco, Los Angeles, San Diego, Phoenix, Tucson, Albuquerque, El Paso, San Antonio, Houston, Dallas, Oklahoma City, Tulsa, Kansas City, Denver, Salt Lake City, and Reno. The Frontier Optronics Network will connect more than 120 metropolitan areas on completion, and it combines advanced fiber with DWDM capabilities, according to the carrier. Like many of the networks, it is also scalable, so that customers can add one information stream at a time as they expand their network capabilities. Frontier has since become part of Global Crossing.

Williams Network, which built national fiber optic networks in both the 1980s and 1990s, is also entering the marketplace. Williams built a national fiber optics network in the 1980s, most of which eventually was sold to WorldCom. However Williams continued to keep two fibers along the network, although for a time it was restricted to transmitting only video over these. However since its noncompete contract with WorldCom has expired, it recently decided to build from that original 11,000 route miles.

Level 3 Communications is planning an extremely ambitious domestic U.S. network, which it anticipates will cost from \$8–\$10 billion to extend fiber optics to cities as well as the surrounding countryside throughout the United States. Begun by James Crowe and the same team that spearheaded MFS Communications, Level 3 had \$4 billion in the bank by mid-1998.

Why was there an opportunity for new carriers to expand the fiber optics infrastructure, since national construction already occurred in the 1980s and DWDM equipment was available to boost those capacities? There are several answers. Perhaps the most compelling is offered by James Crowe, who notes that traditional carriers may not have been so vigilant as they could have been once fiber optic construction began:

Figure 3.4. Latin America has become a very hot throughput for fiber optics traffic, with an increasing number of fiber optic cables passing in and out of the continent. (Map courtesy of Pioneer Consulting)

> The problem is that network operators are wondering how long their networks are going to last, rather than wondering how quickly they need to upgrade to move a bit a mile in a second more cost effectively.... They are still producing product aimed at legacy suppliers to stretch their capabilities, rather than trying to optimize the performance of their networks.

Crowe characterizes what is going on as "an historic shift in networking, one that comes along every century or so."

The Level 3 network will use packet switching, while the fiber optic construction in the 1980s used circuit switches, which are more acclimated to smaller units of information and voice transmission. Level 3 is also Internet-protocol-based, since data traffic is replacing voice traffic (fibertoday.com, 2 July 1998, p. 1).

These 1990s fiber carriers add an entirely new dimension to technology compared to the older carriers by offering new fiber and new optronics/electronics. Deregulation also played a role in creating demand. New carriers, like Level 3, plan to spend $8–10 billion to build networks throughout the United States, including metropolitan areas. This is but one indication that the distinctions between long-distance carriers and local-exchange carriers is disappearing, allowing carriers and savvy businesses to forge their own relationships to achieve their goals.

Many of the distinctions created by Judge Harold Greene when AT&T went through divestiture in 1984 have disappeared, and the rest await only the RBOCs competition in the local exchange as defined by the Telecommunications Act and implemented by the Federal government.

Carriers are also thinking more globally: WorldCom intends to build its own global network, and it has already connected European cities to U.S. cities, largely through its own network. This is an example of liberalization in Europe encouraging U.S.-based carriers to be more forward-looking. In fact while U.S.-based carriers appear to be making some inroads into international markets, the reverse trend generally holds true for vendors.

Japanese and European vendors always played a role in improving fiber optics technology in systems going into the United States; now this is particularly true, especially in the fiber optics arena. Active U.S. vendors in the 1980s, including DSC Communications, Harris, ITT, and Telco Systems, are being replaced by strong international competitors, who sometimes purchase them. Recent examples are Alcatel's acquisition of DSC and Nortel's acquisition of Bay Networks.

The DWDM market appeared more promising with U.S.-based vendors Lucent and Ciena seemingly first and second as the market for eight channels and more evolves.

Recognizing that it would be difficult not only to compete against Lucent but also the other major vendors entering the marketplace, Ciena entered into an agreement to be purchased by Tellabs, which, as mentioned earlier, has done extremely well in the field of DCS for SONET systems. When the Tellabs–Ciena deal fell apart, a number of reasons were suggested, not the least of which was the announcement the morning of the shareholder vote at both companies that AT&T was no longer interested in using Ciena's equipment. That was especially devastating, since a year earlier Ciena had announced that AT&T was examining Ciena's equipment for potential incorporation into its network.

Some industry observers saw Lucent's fingerprints on the AT&T announcement, since it was clear that AT&T had decided to use Lucent's equipment and the two were former partners anyway. However, that has not been proven, and as of this writing, it remains heresay. Regardless of the reason the timing of the announcement was awkward for Ciena and Tellabs—especially for Ciena, which had to restablish its credibility in the marketplace.

Depending on whether Ciena continues in business on its own, and whether other upstart U.S. companies, such as Tellium and Osicom, can make an impact, the DWDM business may very well go the way of the SONET business. Such internationally based vendors as Nortel, Pirelli, and Alcatel are already claiming high-transport capacity systems. Successful entry by Japanese vendors Fujitsu, NEC, and Hitachi could also bring about an international feeding frenzy. DWDM continues in its formative stage as vendors compete for the metropolitan market, although it is moving forward rapidly. Carriers are evaluating these DWDM systems carefully, with an eye to the types of service they can provide.

In these national fiber constructions, carriers battle not only to offer their fiber but also to develop services attractive to customers. Carriers also seek vendors who can provide the most bandwidth at the lowest cost.

Integrating this new equipment can be challenging at times as networks transition from circuit switching to packet switching, from electronic networking to photonic networking, from voice-based systems to IP-based systems. Fortunately in some respects for the optical fiber cable industry, much of what Internet service providers are attempting to do leads to the same optical fiber designs as those used in other fiber cable markets. Thus it is not necessary to redesign the tires when building a new car.

More recently there was a trend among carriers to fit as much optical fiber cable as possible into a duct, and that can include 300 or 400 fibers in a cable. "Optical fiber is becoming a very precious commodity," says Siecor Senior Vice-President Derek Lawrence. When that number of fibers is contained in a cable, mass fusion splicing is necessary. And while losses are higher when using mass fusion splicing than when single-fiber splicing is deployed, such losses "are perfectly acceptable," Lawrence observes.

Cabling companies, such as Siecor, have had to contend with declining price, which receded by a startling 20% or more in 1997–98 alone. "There has been a significant reduction in pricing," says Lawrence. A main driving factor: supply was able to catch up to and exceed demand in that period. Yet these terrestrial carriers, like their counterparts in submarine networking continue to believe that the ever-expanding demand for bandwidth will prevail: People will continue to require ways of accessing more services faster, and they will be willing to pay for that ability. Carriers base this belief on the growth of U.S. data services as more people learn about the Internet, for example, and on the emergence of the Internet in other nations.

Single-mode fiber continues to be the dominant fiber in the industry, perhaps accounting for as much as 95% of the market. Even in premises networks, backbone network cables are primarily single-mode, while multimode fiber is used for distribution, says Lawrence. However it is not at all clear that in the end early steps to bring fiber to the desk will set the pattern for what is to come, he clarifies.

While copper cable proponents used to argue that fiber's size and fragility made it difficult to implement in building-type settings, technicians now view in-building fiber installations very matter of factly: "They splice, test, and install the cable," says Lawrence. "Local electrical contractors can wire the building with fiber for you the same as they would wire it for you with copper or coax." In a return to policies of 20 years earlier, Siecor once again is playing a hands-on role with regard to the cable it is supplying to make sure the plant works. "A lot of our customers have downsized," says Lawrence:

> The telephone companies have had fewer people to help us with. Whereas perhaps 5 or 6 or 7 years ago a customer in the field would do all of the termination and install hardware as necessary, they are looking for us now as to whether that can be provided with a turnkey solution.

Compounding this trend are the number of new carriers: According to one estimate, there were nearly 3000 competitive local exchange carriers and 7500 Internet service providers in 1998. As a result there is more emphasis on prefabricating the system, so that less in-field labor is necessary, says Lawrence. For example Siecor is cabling longer sections to reduce the amount of splicing in the field.

Regarding reliability once the cable is in ground, Lawrence notes that "there aren't many materials as long lasting as glass, glass itself has a tremendous lifetime expectancy." Lawrence continues:

> The cable materials are what determine the potential lifetime. Fortunately, we have ways of doing accelerated aging so that we can project the lifetimes of the materials.

Generally, says Lawrence, the optical fiber cable that was installed was extraordinarily reliable.

Bandwidth itself becomes the primary commodity: If companies have enough excess bandwidth, they will be successful, since it will be used for future applications, or so the reasoning goes. That certainly was the view point for the past several years and with good justification.

What does the future hold? Will we encounter another fiber slump similiar to that from 1987–88? That does not appear likely, at least for some time. Demand now exceeds supply, and prices have begun to rise again. That, as of this writing, factors in major expansion at leading producers Corning and Lucent.

The industry is in incredible shape. For the near-term, Lucent Executive Vice-President Dan Stanzione calculates that the $380 billion telecommunications market will grow to $650 billion by the year 2001. There is little doubt that a growing piece of that enormously expanding pie will be in fiber optics.

4

Fiber Sprouts in Mexico

The only one of the three North American countries that had not taken as full advantage of fiber optics as it had economically made sense to do so in the 1980s was Mexico. Painfully aware of slipping behind its neighbors to the north, and of the pressures placed on its northern neighbors to extend their network capabilities into Mexico to promote their international capabilities, the Mexican government agreed to deregulate its telephone business in the 1990s to bring competition into the market and move things forward.

Deregulation was an important spur for both the telecommunications business in Mexico and for the growth of fiber optics in the country—just as it was and continued to be in the United States and Canada. The unresponsiveness of a slow-moving monopoly in Telefonos de Mexico (Telmex) gave way to a market now enlivened by competition as such carriers as AT&T and MCI strove for continental interconnection. (Note: The Mexican market was and continues to be critical for major long-distance carriers. AT&T's Diane Burness characterizes it, next to Canada, as the largest traffic stream in the world.)

The first part of deregulating the Mexican market went relatively smoothly. Mexico's regulatory arm, Coftel, decreed that the long-distance market was to open formally on 1 January 1997, which in fact occurred. There are now eight carriers in the Mexican long-distance marketplace. One of the carriers, Alestra, a joint venture that includes AT&T, Grupo Alfa, and Bancomer–Visa (which is not to be confused with the credit card company), has taken a 15% share of the long-distance market.

A second venture, Avantel—a joint venture between Grupo Financiero Banamex–Accival and MCI—gained 10% of Mexico's long-distance market share by July 1997.

In all alternative carriers to Telmex took 30% of market share in the first year f deregulation, according to AT&T's Burness. (Note: This was far faster than AT&T gave up market share for example in the United States.)

Powerful networks have brought Mexico into the twenty-first century of telecommunications. The Alestra network, of which AT&T is a 49% owner, installed 4300 km of optical fiber cable by 1998; it includes Lucent's 5ESS switches. Alestra now has more than 1 million operational lines. Probably under AT&T's strong influence, Alestra decided to upgrade its network to DWDM using Lucent equipment. The installation represents Lucent's first DWDM installation in Latin America.

Competing with Alestra is Avantel, the joint venture between MCI and Grupo Financiero Banamex–Accival, Mexico's largest financial institution.

Avantel was the first alternative long-distance carrier allowed to operate in Mexico; it was granted the privilege of supplying traffic in August 1996. Then Mexican President Ernesto Zedillo even made the first call along the network. Avantel provides a complete portfolio of competitive business and residential long-distance, data, value-added, and Internet/intranet services.

Avantel, which has a network infrastructure of 3400 fiber optic route miles that it began constructing in 1995 at a total cost of nearly $1 billion, captured an estimated 10% of the $4 billion long-distance market by July 1997, some 7 months after deregulation.

Another long-distance network is being developed by Bestel, S.A. de C.V., which uses Mexican railroad right of way to install fiber, much as several U.S. carriers built their networks in the 1980s. Bestel is a joint venture between GST Global Telecommunications and Odetel S.A.

The Bestel network was planned for 2270 km, linking 14 Mexican cities and including population-dense Mexico City. Should it so desire Bestel has enough right of way to add 2430 km to its network, more than doubling its size to 5600 km. Bestel has already signed up major U.S. fiber optics carrier Qwest, which contracted for four dark fibers along the Bestel route. Qwest was to be connected to the network at Nuevo Laredo, across the border from Laredo, Texas.

Despite these gains there still are a host of regulatory issues to deal with in both the long-distance and local exchange that impact the very lives of new carriers. "There are still significant hurtles for [these ventures] to be profitable," Burness says.

For one thing access fees continue to be a major concern. At the time of this writing, long-distance carriers paid 58% in surcharges to connect with TelMex's local exchange, which Burness says is a violation of Mexico's commitment to the GAP treaty. The AT&T venture attempted to circumvent this fee by providing fixed wireless service, but it was discouraged by unfriendly regulators.

Another difficulty is that some new carriers end up with "default business," Burness says; that is, Mexican business that is not stable and consequently Telmex decided not to pursue. The implication is that Telmex knew it was not good business but left its competitors to find it out the hard way.

For Alestra, which has already invested some $950 million in the network, the situation has reached a critical stage. As a result Alestra, Avantel, and other carriers with a partial U.S.-based interest attempted to persuade the Federal Communications Commission not to approve a joint venture between Telmex and Sprint in the United States. This effort proved fruitless, since Telmex and Sprint did gain U.S. approval to move forward.

"We need to make sure that we can operate in an environment that supports competition before Telmex reaps the benefits of the open U.S. marketplace," Burness said by way of explaining the AT&T position.

For example Mexican customers do not have the same one-stop shopping in Mexico for carriers competitive with TelMex that TelMex customers have in the United States. Instead, Mexican customers must contact both Telmex and Alestra if they are Alestra customers with a problem. On the other hand, increased competition in the Mexican marketplace due to the first stages of deregulation easily matches what occurred in the United States during a comparable period. Burness and other participants believe Mexican regulators intend to resolve these issues. Mexico has already come a long way, and Burness acknowledges that "the opening of a market is a complex issue."

She is not implying that Mexican officials attempt to make things unnecessarily difficult. Deregulation "started with the best expectations on everybody's part," and she believes the issues are resolvable.

Since the Mexican government has made such important strides in connecting its network to the global fiber optics superhighway, it makes sense to continue until full participation is achieved. However that may not necessarily be achieved in the same way that it was and continues to be achieved in the United States.

5

The Canadian Presence

To understand how Canada gravitated to fiber optics—and why—it helps to understand the particular needs of the country itself. With the second largest national geographical size in the world but a population only about 10% as large as the United States, Canada is acutely aware that it must compete effectively with—and not be consumed by—its larger, more powerful neighbor to the south.

This philosophy produced a culture that in many ways learns from and implemented U.S. patterns of fiber development but on the other developed its own identity by strong collaboration among carriers, vendors, research entities, and the Canadian government. In some respects Canada, like Japan, needs a favorable trade balance to survive economically; thus the high-growth area of telecommunications has taken on elements of a national mission. In fact 60% of Nortel's Canadian workforce depend on international sales. Since Canada has one of the most open, and certainly the most lucrative, telecommunications markets nearby in the United States is a major plus in those efforts.

As reported in *Rewiring*, Northern Telecom (now Nortel Networks) was involved at a very early stage in the development of fiber optics. Canadian carrier Saskatchewan Telephone built a $50 million, 3200-km network that traversed much of that province. Manitoba Telephone was also involved in an early trial (The Rewiring of America "The Fiber Optics Revolution," 1988, p. 44).

Both Canadian and U.S. markets have had national carriers that form the foundation of their telephone marketplace, and both had a central research and vending organization that served as a focal point for efforts. In the United States AT&T was the preeminent carrier before divestiture led to the incorporation of the RBOCs into regional carriers that interacted with AT&T. In Canada Stentor acted as the long-distance arm, and associated RBOCs include Telus (Alberta), Bell Canada (Quebec), Manitoba Telephone (Manitoba), Saskatchewan Telephone (Saskatchewan), and BC Tel (British Columbia). And just as Bell Laboratories served as a major telecommunications R&D arm for the U.S. telecommunications industry (also through Bellcore [now Telcordia Technologies, Inc.] and now embracing elements in both AT&T and Lucent Technologies), Nortel assumed that function in Canada by taking over the functions of Bell Northern Research.

With more than 23,000 employees in 20 business centers throughout Canada, Nortel Networks is Canada's leading R&D spender, accounting for nearly one-quarter of all Canadian R&D expenditures (see Figure 5.1). In fact if there were a Canadian telecommunications monolith around which national telecommunications survival, wherewithal, and prestige were centered, it is the powerful Nortel Networks, which now matches Lucent stride for stride in the North American telecommunications marketplace (see Figures 3.1, 5.2, and 5.3). "Canadian jobs depend on export sales," the company stresses. Nortel has an enormous presence

Figure 5.1. Nortel Networks and its Harlow Labs in the United Kingdom are among the best in the world for studying fiber optic technologies. (Photo courtesy of Nortel Networks)

Figure 5.2. A Nortel Networks engineer works on the Optical Shaved Protection Ring in the Harlow Optical Communications Lab. (Photo courtesy of Nortel Networks)

Figure 5.3. A Nortel Networks employee working at the Integrated Optoelectronic Device Development Facility at the Harlow Labs. (Photo courtesy of Nortel Networks)

in the United States and says it now competes in more than 150 nations throughout the world.

Nortel Networks officials proudly point to their St. Laurent plant near Montreal, which the company characterizes as one of the most advanced production and design facilities in the world. Nortel Networks claims that more than 75% of North American and 50% of European Internet traffic flow through installed Nortel Networks' equipment.

A Nortel strength, many believe, is the company's willingness to partner with other, more specialized vendors: "We do joint R&D with those who want to go beyond a simple vendor relationship," says Philip Bell, Nortel's account vice-president for broadband sales in Canada. "Where there is a missing part, we help to find a solution." Because Nortel assumed the formidable Bell Northern Research, it has laboratories to help it design and develop products for that purpose, especially for Canadian RBOCs.

For example Nortel formed a partnership with Antec to develop voice-over-cable television through subsidiary corporation Arris. While there were various fits and starts among U.S. telecommunications companies entering the cable television market—and vice versa—following passage of the Telecommunications Act of 1996, Nortel and Antec put their heads down and continued to work on the Cornerstone product as part of the Arris project. As a result cable television companies, such as the new AT&T–TCI venture and Cox Communications, are set to use that broader bandwidth to provide the voice services Nortel and Antec engineered. With regard to Cornerstone, Bell notes that there is a market for voice-over-cable television in Canada as well, since cable television passes 98% of the homes in Canada.

When this author had an opportunity to visit the St. Laurent plant in October 1996, Nortel Networks had just begun setting up the OC-192 assembly line. This equipment assembly line later lead the market for 2 years in making OC-192 equipment, which operates at 10 Gbps. Most other fiber optics equipment during that time had a high-end operation of 2.5 Gbps at best.

"The company now is shipping SONET OC-192 product, the first in the world to do so," as *Fiber Optics News* (*FON*, 14 October 1996, p. 2) reported. The first unit, which was sent to WorldCom, initiated a $500 million order.

Nortel also provides BTI Telecom with OC-192 equipment along BTI Telecom's 3250-mile network, which connects New York City with Miami, Atlanta, and Nashville. And IXC Communications, which has since combined with Cincinnati Bell to form Broadwing, decided to use Nortel Networks' OC-192 equipment to replace Alcatel equipment, according to informed sources. Nortel's OC-192 equipment was used in that installation (ftcom 9 July, p. 1).

Nortel also received an important contract to provide its SONET and WDM equipment to British Telecom's efforts to build a pan-European fiber optic network that would span Belgium, France, Germany, Holland, Italy, Switzerland, and the United Kingdom. Nortel will use its 10-Gbps technology as the core of the contract. "We already have achieved quite a presence in Europe, with some $3.5 billion in orders," said Nortel spokesman Thom Hill. "But this job is very important to us. We are going to be connecting six customers in six different nations." (fibertoday.com 9 June, 1998, p. 1)

Due to Nortel's 2-year OC-192 window, the company was able to claim over 90% of the market. Nortel has reached more than $1 billion in sales on that product alone. "There are now 11 networks that are using our 10 Gbps-networking equipment. All of the customers that we have in the interexchange market are deploying it," Nortel spokeswoman Shelley Grandy told fibertoday.com (6/8/98/1). "There is very little competition in the marketplace," she continued. "We are the only ones widely deploying that option."

Using a 16-channel DWDM system, Nortel had the capability to transmit 160

Gbps over a fiber. The vendor reached 320 Gbps shortly thereafter. "This is not just a story about being the leader in providing raw capacity," says Mike Unger, then president of Nortel's Optical Networks Division:

> Horsepower without control and without a guaranteed quality of service is not valuable to a service provider. The winning combination for us has been the ability to deliver capacity along with bandwidth management, full network control, reliability and system performance—and all at the least cost per bit.

Unger says the company is working hard to deliver a terabit on a fiber by the year 2000 and has set the benchmark as a goal the company will reach.

Nortel has also been working hard on the next iteration of the SONET chain, OC-768, which has already been demonstrated in a laboratory setting, according to Serge Melle, director of account marketing for Nortel's transport networks (fibertoday.com, 9 June, 1998, p. 1). That high-end capability with the vendor's introduction of its low-end OC-3 product gives Nortel a complete set of SONET products. "We want to blow the doors off in both directions," said Greg Mumford, then serving as Nortel's vice-president and general manager for SONET networks. "We don't want any holes in the portfolio."

Nortel shipped 27,002 pieces of SONET equipment by June 1996, with 19,200 of those going to the United States, the company reported. That accounted for a global lead, according to the market house Kessler Marketing Intelligence. In fact Nortel captured 36% of the SONET market in 1997, according to KMI, yielding revenues of more than $1.4 billion.

To gain a better understanding of how regulation affected the Canadian marketplace and how fiber optics gained prominence in that marketplace, we must understand the position of Canadian regulators, who have instituted what they call the message minute. The idea stems from the need to be competitive with the United States by developing the best, most efficient, and least expensive tools to transmit information. Thus the rate is set to ensure carriers of deriving the revenues they need to continue growing and competing. Most of this service is sold at a flat rate throughout much of North America.

In comparing various transmission media to deliver the message minute most efficiently, Canadian regulators found that fiber optic technology was "more and more attractive," says Bell, "that's what drove the fiber business." To extend this efficiency, the emphasis was on optimizing the glass in the ground, that is, getting the most efficiencies out of the fiber to reduce costs for all customers and render transmission more cost effective.

The emphasis therefore is on getting the most from available bandwidth. For example there are efforts not to use bandwidth during quiet moments in voice and data transmission, so that it can be used for something else: "If we stop talking on a data-facilitated business, the silent moments are not transmitted," says Bell. "We are eliminating those silent moments to save bandwidth." The reason? "Canadian demographics require us to communicate more effectively."

Canadians are proud that their Internet use equals that of the U.S.—and may well be ahead. Acknowledging that Canada is less challenged by immigration and a larger population, Bell believes Canadians are generally more learned than their neighbors to the south; six of every ten homes in Canada has a computer, for example, says Bell. Whether it is due to the international business its Nortel Networks has already developed or the market impact that deregulation has engendered, Canadian regulators have opened up the domestic telecommunications market, and the resulting frenzy is similar in some respects to what is going on in the United States. For example the number of long-distance carriers in Canada increased to four. The traditional carrier, Stentor, also may be getting competition from Telus, one of the four RBOCs supporting it. Telus plans to build its own long-distance network servicing eastern Canada.

Fonorola, a recent long-distance carrier with substantial fiber right-of-way in both the United States and Canada, was recently purchased by Call Net, which has ties to Sprint. The plot thickens even more as Bell Canada purchased Ledcor, a construction company that is building its own national fiber optic carrier's carriers network in an agreement with the railways. AT&T long distance has been in business for a number of years in Canada, as well.

Then there is Metronet, which strives to be the first and largest national local exchange carrier in Canada. In fact there are a number of competitive local exchange carriers, some of which have been selling data services since data was deregulated. There also are 32 cable television companies in Canada, a number of which have been selling data services for some time. In fact many of the cable television companies have been offering data as a primary service. A substantial number use hybrid fiber-coax architectures.

Bell acknowledges that satellite is still more a competitor technology in Canada than it is in the United States, particularly because of Canada's rural demographics. A primary deployment is direct television to home services.

When fiber optics made strong headway economically into Canadian networks, proponents felt that fiber to the home was the only answer. This movement was slowed here and around the globe by economics. However Bell believes that the strategy is starting to make sense in fiber-to-the-multiple business unit and multiple-dwelling-unit building. "That's because the residents are using more bandwidth," he observes.

What is on the horizon for Canadian fiber? Gigabit opportunities throughout the network will play an important role, Bell says, particularly as web-tone networks come into being. Data-oriented networking is becoming critical, and Canada hopes to be right in the middle of it. Bell notes that Nortel took a "right turn" when it realized the importance of data, and now through the acquisition of Bay Networks and other factors, Nortel is ensconced in that effort. In fact Nortel's acquisition of Bay Networks represented a tacit admission that perhaps Nortel had not gone so far as required in adjusting to data communications. Bay Networks provides products to serve corporate enterprises, service providers, and telecommunications carriers. It offers Ethernet and ATM LAN switches, routers, shared media, remote and Internet access solutions, IP services, and network management applications.

To demonstrate how important this acquisition is to Nortel Networks, it established an office of the CEO to be shared by the CEOs of both Nortel and Bay Networks in making corporate decisions. Bay Networks' CEO did leave shortly following the acquisition, however.

Nortel is also using its clout to take positions in other companies. For example it initially took a 33% equity position in Photonic Technologies, based in Sydney, Australia. "We saw they have a very good technology and wanted to give them the tools they need to expand their customer base and grow their own business," said Steve Turley, director of strategic business alliances for Nortel's optoelectronics division (fibertoday.com, 3 May, 1998, p. 1). It eventually bought the company.

Nortel's optoelectronics division will provide specialist engineering and manufacturing for Photonic Technologies' circulator products. "There is a rapidly increasing global demand for high-quality, high performance, cost-effective optical circulators and filters," according to Ralph Betts, managing director and cofounder of Photonic Technologies. "This investment will enable Photonic Technologies to offer these components in volume to meet this demand."

More than 85% of Photonic Technologies' revenues now accrue from exports to North America, Europe, and Asia.

Nortel Networks also purchased Cambrian Systems, an acquisition that could be worth $300 million. Nortel views this purchase as the critical link in bringing its DWDM capabilities into the local exchange a year earlier than if Nortel had developed the technology on its own. The purchase also responds to industrywide criticism that Nortel's 10-Gbps-based technology does not make sense for adding and dropping wavelengths in the local environment, at least as it is currently configured. However even before the ink dried on the contract, industry observers were wondering about the wisdom of Nortel Networks spending $300 million on a company that—at the time of the acquisition—did not have customers. Nortel and Cambrian officials acknowledged that Cambrian brought only a mind share, but both believe that will change shortly.

Nortel Networks is acutely aware that Cambrian lacked customers at the time of the acquisition. In fact in their agreement, Cambrian has a floor of $240 million, with the potential for employees to gain $60 million if they reached revenue, market share, and new-customer goals.

Bringing Cambrian into the Nortel mix seems to be a laudable goal. Nortel's OC-192 DWDM system leads the North American interexchange market, but there are problems getting that data flow into the local exchange. "There is no doubt there is a bottleneck between the enterprise network and local exchange," said Nortel's Brian McFadden, vice-president and general manager of optical network applications. "That is what we are trying to create with this, a way to overcome that. We are creating a new market demand and paradigm." One obvious advantage for Cambrian is Nortel's marketing influence which should bring Cambrian networks that it had not penetrated.

A buzzword on the tips of many Nortel tongues these days is optical Internet, and the carrier earnestly believes that Cambrian's 32-channel protected and 64-channel unprotected metropolitan DWDM system can help it reach that goal: "Cambrian is very important in our completing the optical Internet," said Nortel CEO and vice-chairman John Roth. Nortel Networks plans to accelerate deployment of the optical Internet deeper into the network by extending its leadership into the metropolitan area network. This will benefit enterprise customers who experience bandwidth shortage, service delays, and costs associated with protocol conversions. Nortel says that more than 75% of Internet backbone traffic in North America is carried over its optical systems. "Despite our long-distance gains, we have not had a big marketshare in cities," acknowledged Roth.

Newbridge Networks, which is selling Cambrian to Nortel, will continue to have a stake in Cambrian and act as a channel for selling its products. McFadden said that Newbridge switching and network manager equipment will continue in at least some manifestations of the Cambrian product line. In today's marketplace, McFadden said that such ongoing relationships with Newbridge are not unusual, and he labels the arrangement "coopetition." Such coopetition assures Newbridge and associates that they are not too closely tied to Nortel. Said McFadden:

> In order to be competitive, you have to come in at less than OC-48.... The first wavelength may be OC-3, then OC-12. The question is what are the next 30 wavelengths going to be?

Nortel officials promised to keep the 170-member Cambrian team intact, and in fact Nortel added some 40 Nortel designers to Cambrian's Kanata, Ontario, plant to try to complete the Cambrian design.

Nortel has been patient with the U.S. utility marketplace as these utilities become more savvy in providing telecommunications infrastructure. The list of utilities using Nortel's SONET equipment is growing rapidly, and it now includes such major players an DukeNet Communications, Tacoma Public Utilities, MP Telecom, TeCom, Inc., and Whisper Communications. This too is a growing area, and Nortel will be central to its optical provisioning.

As the global fiber superhighway continues to expand, we expect the Canadian network to be connected to it and an advanced part of it. Stentor and MCI recently agreed to interconnect, and some transoceanic pipes terminate in Canada.

Canada understands the importance of a sound, fiber optic-based infrastructure, a philosophy that will serve its economy and people well in years to come.

6

Bandwidth as Precious Commodity

The eminent futurist George Gilder posited that the ability to provide bandwidth in and of itself may be a communications carrier's most important asset. This represents a new way somewhat of considering the telecommunications world, where bandwidth is a valuable currency for the new millennium.

As we know new organizations often have an inherent advantage, since they are starting with the latest technology, unencumbered by older methods or processes. For example the destruction that Japan experienced during World War II forced it to rebuild a new infrastructure that in large part catapulted Japan among the leaders of the industrialized world.

Rewiring lauded such carriers as AT&T, Sprint, and MCI for their willingness to base their nationwide infrastructures on the new technology of fiber optics, yet now, 10 years later, these are the traditional carriers. New carriers, such as WorldCom, Qwest, the reborn Williams Networks, IXC Communications, Level 3 Communications, and Frontier Corp. are the innovators. (No doubt because of their increasing value, IXC has since been purchased by Cincinnati Bell to form Broadwing, while Frontier was purchased by Global Crossing.)

What has happened to cause this changing of the guard? This is due to the rapid growth of fiber optics technology, which was taking place even while these first fiber optic networks were establishing a national presence. Some of these carriers attempted to build the new technologies into their networks; others have simply been at their mercy.

James Crowe, the eminent fiber optic technologist, engineer, and network leader, by overseeing the initial construction of the MFS network, and now building Level 3 Communications (both innovative fiber constructions of their time), has been around for both decades of fiber growth and has some prescient thoughts on the matter.

The capabilities of routing and associated technologies are reaching an order of magnitude improvement every 13 months or so, while transmission capabilities are also expanding thanks to DWDM and related optical technologies, says Crowe. Yet carriers traditionally are only building their networks every 13 years or so. "That's a long time," says Crowe, "it helps to explain why so many new entrants can get into the business over more established carriers." He adds

> The problem is that network operators are wondering how long their networks are going to last, rather than wondering how quickly they need to upgrade to move a bit a mile in a second more cost effectively.... They are still producing product aimed at legacy suppliers to stretch their capabilities, rather than trying to optimize the performance of their networks.... It's a little like Intel continuing to build 286s and 386s because that is what some of their customers still want. (fibertoday.com, 2 July, 1998, p. 1)

It is not difficult to see the irony here: The interoperability so critical to the success of these earlier carriers in bringing fiber into a voice-oriented environment in some respects is now an albatross around their necks, particularly as newer carriers introduce more modern systems. More specifically by not continuing to introduce the newest fiber optic technologies into their networks, unlike the proverbial chameleon that can grow out of its old skins, earlier carriers are saddled with an infrastructure that runs the risk of no longer being able to keep up or of being directed at a decreasing market, voice, rather than data, whose demands now drive most of the growth.

As Crowe explains, "There has been an historic shift in networking, one that comes along every century or so." This shift was driven by the need for more bandwidth and, specifically, for more data networking. The majority of our communications networks are becoming data-driven (Internet and related services) rather than minority voice-driven. This change profoundly impacts how these networks are now designed, since traditional voice-centered networks deployed circuit switches, which are good at transmitting voice and poor at sending data. With the advent of data, newer, faster, better networks using more efficient packet switching and services make more sense.

While this situation is not particularly good for the first national fiber optic network builders, who invested billions of dollars in their networks and whose networks are populated with circuit switches, it represents an important advantage for newer carriers who can deploy these new protocols. What does this mean for vendors who are trying to keep up? Cheaper, faster, better, and newer are important considerations. When Ciena had the glamorous technology of 16-channel DWDM, new carriers flocked to it because it increased bandwidth. However these carriers sometimes believed they were not getting the margins per channel that they thought they should when compared to installing more fiber throughout the routes. A natural part of the curve as bandwidth becomes more ubiquitous is that it will also have to become more affordable. Carriers are not going to pay for technology alone; they will pay for the improved economics that new technology brings.

Do vendors attempt to make the most money possible from each order when they have a golden window on a new product that delivers more bandwidth, or do they reduce the cost, thereby giving carriers a true break in hope of attracting more long-lasting contracts that will continue for the longevity of the product? There are indications that at least one of the early DWDM vendors attempted to make too much money in too short a period of time. While it was great for the company's executive crew to point to the company's profits, the company sacrificed a major portion of its larger business by doing business this way. The situation is reminiscent of the old Wall Street adage that bulls win, bears win, but pigs never win.

Nortel on the other hand seemed to adopt a price scheme that encouraged customers to stay with it regarding its OC-192 equipment. While there were some early complaints about the price of the equipment, Nortel was able to make major inroads and shore up business even after several of its competitors also reached OC-192 capability.

The cost of bandwidth will decline; it is simply a matter of economics. Vendors must recognize that they also contribute to the goal of facilitating communication. Vendors who provide better products and pass on the lower costs that increase usage; the others will not.

Such carriers as Qwest and WorldCom were able to install the latest generation of optical fiber, linked to the fastest lasers and using the most advanced DWDM equipment to obtain maximum bandwidth in a system. They are building this equipment with the firm belief that needs will increase and customers will inevitably appear. In short these carriers place bandwidth at the top of their list of priorities, and they assume that people and businesses will find the wherewithal to use it. To date these carriers have been phenomenally successful; Qwest shows a profit months before its capital-intensive fiber optic construction is completed.

In many instances these newer carriers took new business from older, more burdened networks, because of their capabilities. in many respects they simply exercised wise old business maxims: Stay ahead of the customer and keep the customer satisfied. Another economic factor bears mentioning: By providing more bandwidth, new carriers are driving down the cost of bandwidth, which has also hurt more traditional carriers, who now recognize that individual message units, such as phone calls, will not always provide the same steady revenues. Bandwidth is no longer something that is doled out on a piecemeal basis; it is available in large amounts for those who have the creativity to use it effectively.

Considering this situation why are some of the more traditional carriers not going out of business? The most direct answer is that the entire pie has grown larger. Even though such data services as the Internet have experienced exponential growth spurts at times, voice traffic has also expanded, albeit at a far slower rate. In short there has generally been enough for everyone, at least to this point.

It would also be premature to assume that these older carriers are no longer active competitors in the bandwidth business: AT&T for example signed a major contract with @Home, the national Internet provider that a growing number of cable television companies are using. The AT&T accomplished this by promising a national DWDM over IP capability over its huge 51,000-mile national fiber optic network. Sprint is also developing the novel ION networking scheme, which promises a more powerful and responsive network over its entrenched fiber optics base.

How can emerging fiber carriers ensure that what is happening to older network operators does not happen to them 10 years from now? Solutions that Crowe's Level 3 attempted to deploy include all-packet switching, all-fiber connect, and duct space to accommodate new generations of fiber not yet developed. Most carriers today have one conduit available for fiber, or at most two, Crowe observes; "we are going to make sure we have between six and eight." He adds

> We are already on our third- or fourth-generation of fiber, depending on how you look at it. We are going to have to be able to accommodate generations five through nine.

The idea that amount of bandwidth rather than another carrier feature is the reason for all of the new high-end construction, including the Oxygen project. And the truth is, no prime-time network has yet lost money by betting on more bandwidth.

Many years ago one of the founding fathers of fiber optics, Charles Kao, observed that one day people would use bandwidth as we now use water. Every indication in our global *infobahn* society tends to support Kao's predictions. as a result of fiber optics, how we look at bandwidth is changing dramatically. For verification consider WorldCom's financials prior to acquiring MCI. The carrier had a net income of $193 million for the first quarter of 1998, and that was after acquisition charges, which included the retirement of debt! Net income the year before was $25 million.

Where has putting bandwidth first led WorldCom? It moved the Mississippi-based carrier onto the global scene, with WorldCom now stringing fiber to the leading cities in Europe, as well as to the Far East. As of 1998 the carrier had the capability to connect end-to-end more than 4000 buildings in Europe with more than 27,000 buildings in the United States—all over high-capacity circuits. The company now operates in 50 nations.

Let us look at the acquisition of MCI. WorldCom is not larger than MCI, yet placing bandwidth first moved it into a superior position. The MCI attitude in some respects was not to revere bandwidth but to derive as much from existing fiber as possible. WorldCom's more progressive philosophy will permit it to acquire another carrier far older and more experienced than itself.

WorldCom President and CEO Bernie Ebbers noted

> A year from now—due to the rapid growth rates in data, Internet, international and local services—MCI WorldCom will be a company comparable in size to many of the more familiar "blue chip names" but with growth rates and operating margins that should distance us from our traditional peer group.... This is an outlook we can all be excited about.

Section 3: The Far East

Asia is a marketplace alive with fiber optics. From the leading-edge role of the Japanese to the committed Malaysians to the open-market-oriented Australians to the booming Chinese telecommunications networking, fiber is well-known and well-liked in the Far East. The Japanese led a remarkably concentrated effort in the area of optical communications from the late 1970s. While seminal discoveries initially came from the United Kingdom and then mainly from the United States, the Japanese entered the field when commercial scale-up became a real possibility, and they have not left.

The Japanese impact around the world—but particularly in the United States and the Far East—was profound. For example when AT&T was looking for vendors to supply 90-Mbps equipment to its first long-distance fiber optics network—the Northeast Corridor project—in the early 1980s, Fujitsu and NEC offered fiber optic electronics that could reach a greater—to 400 Mbps (*ROA*, p. 36).

While the first national U.S. fiber optic network was completed in 1985, the Japanese had a national fiber optic network operating in 1983. In fact the ability of Japanese companies to make diode lasers led them to dominate the compact disk (CD) field. The consumer appeal of CDs yielded a much larger market response than fiber optics, reaching a marketshare substantially over $40 billion. With their expertise in transmitters and receivers, Japanese companies also took a leading role in SONET/SDH development, as well.

Japanese companies also make substantial amounts of optical fiber; in fact several key Japanese fiber makers operated in the past as licensees of Corning. (Note: Through major fiber consumer Nippon Telegraph and Telephone Public Corp. [NTT], Japan unfairly excluded Corning from its own market almost from the time that Corning began making fiber and illegally established a factory in the United States—Sumitomo.) Corning was helped by the U.S. International Trade Commission (ITC) in shutting down Sumitomo's North Carolina plant when the ITC ruled that Sumitomo's optical fiber manufacturing processes clearly violated those of Corning. Since those disturbing times, Corning was able to penetrate the

Japanese market to some extent, although much of its success was with alternative carriers that are emerging, and not with NTT and its international arm, KDD.

Regarding the Japanese market, Siecor Senior Vice-President Derek Lawrence notes, "There is very significant growth ongoing there. The growth in Japan has been comparable to growth in the United States." While NTT represented some 65–70% of the total fiber market in Japan in 1998, Siecor also made inroads selling to other customers there as well: "We also sell to the power companies and the new common carriers," says Lawrence.

NTT was the leading domestic provider of fiber optics networking in Japan, although the nation was farsighted enough to allow several new common carriers (NCCs) to enter the Japanese market around the time of AT&T's divestiture in the United States. Each of these three NCCs built competitive fiber optic networks in Japan, but as of 1998 none had taken appreciable marketshare from NTT.

The NTT also led the globe in installing extremely high-density optical fiber cable to outfit suitably its densely located, highly productive workforce. The megacarrier uses single cables provided by Siecor and Japanese cablers that contains up to 1000 optical fibers. The NTT also tested single cable that houses the 3000 optical fibers (*FON*, 5 February 1996, p. 1). (Note: Deregulation also allowed KDD to install its own terrestrial fiber and NTT to enter the international submarine fiber optic cable business in competition with KDD.)

Australia recognized the importance of using fiber optics relatively early and therefore established two competing national long-distance carriers. The result was the installation of fiber throughout Australia.

An important fiber optics user in the 1990s was the Peoples Republic of China, which also adopted a competitive long-distance structure. (Note: Fiber optics is also being used throughout many Chinese provinces.) Lucent Technologies made a special commitment to China, which may serve the vendor well as China has the potential to be the largest user of fiber optics networking in the world after the millennium.

South Korea, which also has a Lucent connection, made optical fiber for more than a decade, and it is spreading cable throughout its nation.

India is establishing fiber-making facilities and upgrading its telecommunications infrastructure with fiber. Lucent entered into a joint venture with Finolex to make optical fiber cable there.

Since the growth of optical fiber bypasses political boundaries, Vietnam has become a major user, developing 3000 km of fiber with Nortel's help; U.S. companies are among the vendors.

Inner Mongolia installed an enormous SDH network across the breadth of the country, running to 7208 km. Israeli-based ECI Telecom is a major supplier of that construction.

Malaysia and Indonesia also installed fiber optic networking, and submarine supertrunks connect Japan, Australia, and China with the global fiber optic superhighway and their neighboring nations, as well.

In short the Far East is extremely well-connected to the world. In fact language, custom, and nationality no longer separate nations, and fiber optics is responsible for removing physical barriers to communication.

7

Japan's Twenty-First Century Infocommunications Superhighway

In the 1970s the Japanese correctly spotlighted optical communications as an industry of the future. They focused thousands of the best engineers in their nation on fiber optics and laser development. Japan mobilized the issue into a national referendum and awakened its population of 125 million, 77.4% of whom are urban-based. This strategy, which works in a largely homogeneous society, such as that of Japan, paid enormous dividends.

One result was a substantial global lead in fiber optics electronics, as manifested by such vendors as Fujitsu and Nippon Electric Corp. (Note: The term fiber optic electronics in some respects is confusing: It actually includes the photonics equipment that operates the system, such as lasers, receivers, multiplexing equipment, and so on, but it does not include fiber, cable, connectors, or electron-producing equipmnent of non-fiber-based systems.) Japanese companies also took a major position in what became an extremely lucrative market—compact disk development. This was a direct offshoot of their laser research.

This concentrated effort also led to the first high-speed electronics for fiber optic systems, according to Takashi Touge, senior manager of the optical network systems division of Fujitsu's telecommunications systems group. According to Touge:

> By the beginning of the 1980s, we had developed 400 Mbps and 1.6 Gbps fiber optic electronics for NTT and other carriers.... The 400-Mbps systems were for the first commercial single-mode fiber systems.

The NTT began building their nationwide fiber optic system from Hokkaido to Kyushu in 1981, and it was operational by 1983, says Touge. Electronics were provided by Fujitsu and NEC. The NTT upgraded the electronics 2 years later to the then astounding rate of 1.6 Gbps. This implementation was also national in scope, Touge says. Much credit should be given to Japanese engineering for accomplishing such a feat; however it should also be noted that with its highly productive and concentrated population, the Japanese actually needed high-speed transmission at an early stage.

"In the middle of the 1980s, we started to study new synchronous networks," recalls Touge. As mentioned earlier Fujitsu was directly involved as CCITT developed SDH standards. "We started to provide this equipment to NTT in 1990," he recalls. There were actually three kinds of equipment; these included SONET, SDH for the Japanese market, and SDH for the ETSI market serving China and some other countries.

Unlike the United States, in Japan there was no real differentiation between the local exchange and the long-distance market, says Touge; NTT simply built full networks that encompassed both. Yet as recorded earlier Fujitsu decided to service the local exchange in the United States, and not the long-distance market, because that was where it saw the greatest opportunity in the United States, Touge says. Much credit should be given to telecommunications labs for developing fiber optics technology, including Fujitsu labs, says Touge.

The year after NTT's 400-Mbps nationwide fiber optic network was operational, the Japanese government allowed the three NCCs to form and compete with the telecommunications monopoly. Given NTT's name recognition and its powerful network, this was not an easy task. One NCC was DDI, which was established in 1984 and granted a Type I common carrier license in 1985. After launching regional cellular services in 1989, DDI also entered the fiber optics business. A separate NCC, the Japan Telecom Company Ltd., is partially owned by the Japan Railway Group, which provided Japan Telecom with its railway right-of-way to lay optical fiber cable. Japan Telecom became public in 1996, then merged with ITJ, Japan's second largest international carrier. It is using Ciena DWDM equipment to expand capacity along some of its more traveled routes.

The third NCC was Teleway Japan, which began providing switched telephone service in 1987. Because it is partly owned by Japan Public Highway, Teleway Japan was allowed to lay optical fiber cable along expressway rights-of-way. Teleway Japan, a consortium of 330 companies led by Toyota Motors, links its success to that of fiber optics. Teleway's fully optical network was completed in March 1995.

According to a report developed by the Japanese Technology Evaluation Center (JTEC):

> Japan now dominates some 90% of the world's optoelectronics markets, and can be expected to continue its dominance for a number of years.... The current size of the Japanese optoelectronic industry is $40 billion; that of the United States is $6 billion. Obviously, Japan has had enormous success with its development strategies for optoelectronics.

As stated earlier the main reason for the success is Japan's inroads into the compact disk marketplace.

In the report Bellcore's Paul Shumate, a highly respected fiber optics pioneer, observes that due largely to NTT's driving influence:

> Japanese transmission equipment pioneered in the use of single-mode fiber, high bit rates and long wavelengths, and NTT was early to commit to the ATM standard for multiplexing and switching.

These characteristics put Japan well ahead of its competitors.

In effect the Japanese severely rattled the foundations of the Corning–AT&T fiber optics infrastructure created through the original cross-licensing agreements that represented the cornerstone of the industry. In electronics the Japanese became leaders in delivering high bandwidth through fast lasers and advanced SONET and SDH systems.

In SONET Fujitsu reached a level of prominence with AT&T in providing equipment in the United States that accounted for a 33% share, according to the respected market research firm of Ryan, Hankin, Kent. However the Japanese were stopped in optical fiber production by Corning's patents. While Japanese licensees of Corning's flourished, Sumitomo, which claimed to have a process that differed from Corning's, was limited in the U.S. and other markets where Corning claimed preeminence.

If there were a company willing to fight as hard as necessary to preserve the cross-licensing agreement, it was Corning, not AT&T. (In fairness AT&T was still restricted by telecommunications regulatory considerations.) As a result Corning, even more than Lucent, became the global leader in providing optical fiber. And Corning fought for its patents throughout the world. For example in 1998, it brought a second legal action against Plasma Optical Fibre, B.V., a Dutch fiber producer that Corning accused of infringing its patents; Corning won the case convincingly. Corning's defense of its patents also stifled illegal competition in other nations. In Canada for example Nortel sold its optical-fiber-producing capabilities under threat of a potential lawsuit, and it no longer has a market in optical fiber.

This demonstrates the depth and power of the Japanese effort, which competed with U.S. and other vendors for every inch of the fiber optics marketplace. Hitachi, Fujitsu, NEC, Anritsu, and Ando all become an everyday part of the global fiber optics landscape.

Japanese vendors even began developing laboratory experiments that were once almost entirely the domain of Bell Laboratories. The NEC for example set a record for transmitting data over a single fiber at 2.64 terabits per second (*FON*, 7 October 1996, p. 1). This followed announcements that AT&T, Fujitsu, and NTT had broken the 1-terabit-per-second mark.

While Japanese vendors were advancing into international markets, particularly in the United States, U.S. and other international vendors were often having a difficult time grasping NTT's procurement tracks and the general procedure for applying for contracts. Although such companies as AT&T and Corning received some work following substantial investments in NTT, evidence showed that the Japanese were in no way giving as good as they were getting. Certainly one reason for this was oftentimes the superiority—or at least strong competitiveness—of Japanese products. But as often chronicled the Japanese have a more homogeneous culture than do many other societies, for better or worse this national trait came into play as fiber optic markets developed.

To be fair the Japanese have taken some steps (however pro forma) to open their markets. For one thing NTT held workshops in the United States and elsewhere to encourage international vendors to bid on NTT contracts. For another the Japanese government allowed the new common carriers to compete with NTT, although they had a formidable challenge, to say the least.

Japan's Ministry of Posts and Telecommunications (MPT) has clearly been at loggerheads with NTT. The question is whether Japan will fully open its market, moving from a greenhouse to an open-air market as the U.S. Federal Communications Commission's Peter Cowhey and Emily Murase describe it in a White Paper published in 1997:

> The relentless winds of the Information Revolution are battering the fragile walls of the greenhouse.... (Greenhouse to an open air market: Transition in the Japanese Telecommunications market.) Japan has important decisions to make. Either it can reinforce the walls of the greenhouse and otherwise attempt to discipline the forces of nature, or it can accommodate the winds of change by bringing down those walls to create an open air market, offering not a single product, but a variety of goods that have benefited from random cross-pollination with external elements.

The paper acknowledges Japan's effort to deregulate by creating the three alternative carriers, NCC. The MPT also created two new common carriers to compete in the international market against KDD. These were International Telecom Japan (ITJ), which later merged with Japan Telecom, and International Digital Communications. The limit to international ownership in any of these ventures is 33%. The IDC has the highest rate of foreign ownership, with Cable & Wireless possessing an 18% share.

The MPT role in taking on fully entrenched Japanese bureaucracies is not an easy one. It must not only match wits with NTT, it must also do battle with the Ministry of Trade and Industry (MITI). The MITI contested MPT's regulatory authority over value-added networks, citing its authority to regulate computer equipment suppliers and manufacturers. However MPT continues as the principal regulator of the telecommunications sector. Cowhey and Murase acknowledge that these reforms have helped, citing increased competition, reduced rates, and new services. As of 1997 123 companies provided facilities-based services and over 2800 resellers operated in the Japanese market.

The authors note that the price of long-distance calls in Japan has plummeted. For example Japanese consumers paid roughly $4 for a 3-minute daytime call between Tokyo and Osaka in 1985. In 1993 NCC subscribers paid only $1.70, and NTT subscribers paid $1.80. Despite the progress Cowhey and Murase note that NTT still controls 93% of all calls. And unlike dominant carriers in the United States, according to their research NTT can use revenue from its monopoly in local calling to subsidize lower rates in domestic long-distance calling to undercut its competitors:

> In 1988, the MPT authorized TTNet, an affiliate of the Tokyo Electric Power Company, to operate the only local network for telephony in the Tokyo metropolitan area in direct competition with NTT.

However TTNet subscribers were required to use a numbering sequence quite distinct from the conventional one, which the FCC report calls "clearly a structural barrier to competition." Given "unequal access," or "dialing asymmetry," the report notes that it is not surprising that TTNet generated only $25 million in revenues by 1993.

On the international front Japan's international arm, KDD, had strong control over the submarine fiber optic networks in the region, and it doled out business to preferred Japanese vendors, much as AT&T did in the United States.

The NTT and KDD roles also changed as a result of deregulation: NTT became active internationally, planning submarine routes connecting key points in and to Japan. The Cowhey–Murase report finds that international carriers have eroded KDD's share from 100 to 75% in the first 7 years since deregulation. The AT&T lost 41% during the same period, the report says. The price of a call from Japan to Washington, D.C. equaled a call in the opposite direction, however, the authors note.

The NTT is also attempting to thwart Japanese government attempts to open the market further by rejecting government sanctions designed to encourage alternative carriers in favor of voluntary measures. In fact NTT called for a 3-year postponement of structural reform to permit new voluntary measures to take effect, the report notes.

Cowhey and Murase recommend several steps to level the Japanese playing field: A truly independent regulator, an open and transparent regulatory process, and competitive safeguards for interconnection. With regard to this third recommendation, the domestic NCCs were frustrated for 5 years in attempting to achieve interconnection with NTT. They finally went to MPT, which told NTT to interconnect with them. The NCCs were then forced to engage in interconnection negotiations with NTT on a service-by-service basis. The MPT finally developed overall interconnection guidelines for interconnection based on a proposal NTT submitted in 1995. Without Japan's homogeneous cultural characteristics, it is highly doubtful that it could even begin to propose a national plan for fiber optics-to-the-home construction on a national level.

The NTT in 1990 clearly was thinking of potential advantages when speaking of fiber optics and its national direction. In a paper entitled "A Service Vision for the Twenty-First Century: Realization of Visual, Intelligent, and Personal Communications Services," NTT observes, "In the twenty-first century, the world is expected to evolve towards more interdependent and peaceful societies." For this purpose active communications is indispensable, and if telecommunications is to play a major role to this end, a clear-cut vision of its future course is necessary. Hence by envisioning 15 years from now (the year 2005), NTT is setting its service target to provision visual, intelligent, and personal services.

The paper refers to an information-intensive society, one that will account for a growing part of the national GNP:

> Information will play a more important role as a management resource. Thus, the production of valuable information, its efficient transmission and effective utilization, will all be required more than at present.... With a single optical fiber, 10 million telephone circuits and 10,000 image circuits can be handled at one time.... Moreover, mobile communications that permit the usage of 100 million portable "pocket phones" throughout the country will become possible.

A follow-up NTT paper in 1994 continued to sketch out details of the infocommunications society. Noting once again challenges of Japanese society, including an aging population, an overly concentrated citizenry on Tokyo, and the need for the free flow of international traffic, NTT foresaw the transition to an intellectually creative society based on infocommunications. As the paper observes:

> We find it difficult to solve these various problems using the conventional methods of industrialized society, based on the movement of people and goods and the consumption of vast amounts of energy.... In the intellectually creative society based on high performance infocommunications of the twenty-first century, great importance will be attached to the free creation, circulation, and sharing of information and knowledge as social and economic assets.

As the result NTT introduced a plan for completing a nationwide fiber optics network by the year 2010. By the year 2000 for example NTT was embarking on an aggressive program to wire all large business in major metropolitan cities with fiber.

Recognizing that the cost would be high, the government proposed including subscriber optical fiber cables in tax measures designed to encourage the installation of a next-generation communications network. As a further incentive, conditions governing the use of roads should be eased, fees for constructing overhead lines along roads should be applied, and the government should construct and improve conduits, lines, and pipes, the paper says. The paper also encourages the establishment of a Giga Bit Network council and Teletopia project and recommends connecting research institutions to high-speed networks.

At about the same time Japan's Ministry of Posts and Telecommunications issued a paper noting:

> As it approaches the twenty-first century, Japan faces many problems, including dealing with an aging population and sustaining economic growth.... To overcome these issues, Japan must conquer the limitations of an industrial society and reform its socio-economic system.

The paper, which also encourages the construction of a national fiber optic network by the year 2010, sets out a timetable. By 2000 the main districts of all prefectural capitals and a portion of the cities designated by the government as "teletopia cities" are to be wired; this means that 20% of the population would have direct fiber capability. By the year 2005 all cities with a populations of 100,000 or more would be connected. The entire population would be served by the year 2010.

The U.S. companies were quick to realize the potential benefit to U.S. fiber optic and high-technology companies. By one estimate U.S. companies could receive as much as one-fifth of the business of building the fiber optic network and selling services on it. (International Trade Reporter, 1994, the Bureau of National Affairs, Washington, D.C., "Japan's Planned Information Highway May Yield Big Opportunities for U.S.")

Not surprisingly NTT is considered the most enthusiastic supporter of the project, although the broader Japanese community was also involved. Key meetings among Japanese business titans were held to enlist support. The NTT would obviously have to remove its barriers to international suppliers if it were given a lead role in constructing such a network, thereby accelerating a change that has already begun. As one knowledgeable official told us:

> There are two ways to get NTT contracts: If they really want the technology, they will buy it outright and quickly. The more traditional way is the laborious, tiered structure.

Perhaps NTT's need for larger amounts of supplies will lead to more reliance on the first means, or at least the number of contracts awarded to international firms will be substantially larger.

Such U.S. suppliers as Lucent, Corning, and CommScope are already jockeying for position. Through subsidiary AT&T Japan Ltd., AT&T has been working with NTT to provide fiber to the subscriber loop. Lucent provided its 5ESS-2000 digital switch and other sophisticated network equipment. Before the creation of Lucent as a separate company, AT&T also established AT&T Yazaki Fiber-Optic Cable Company Ltd., a joint venture with Yazaki Electric Wire Cable.

Lucent has established a laboratory near Tokyo to study passive optical networking technology, which is expected to be a key enabler as fiber migrates to the office and home. The buildup is directly related to Lucent being selected by NTT to provide Passive Optical Networking for NTT's nationwide construction. Lucent is expected to select as many as 69 technology experts in Japan to staff the facility by the year 2000.

If the truth be told U.S. companies like Corning and Commscope have had much more success with the alternative Japanese carriers. "We have been finding more success in the non-NTT market than in the NTT market," Corning's Tim Regan has observed (*PS*, Nov 1994 p. 66).

Other alternative carriers in Japan, such as tiny Tokyu Cable are also trying to carve a niche. Tokyo Cable, which uses right-of-way provided by its parent company Tokyu Corp., is building a largely fiber cable television infrastructure.

Commscope's Michael Ellis, director of regional sales for Asia and the Pacific Rim, found that "the situation regarding providing telco equipment to Japan is improving significantly." Ellis notes that while fiber used to run to nodes of some 2000 residences, it is now down to 500. "The node sizes are getting smaller and smaller, a sign that fiber already is coming closer into the neighborhood," Ellis observes.

The NTT was involved with U.S.-based Silicon Graphics since 1994, a relationship through which the Fortune 500 company is expected to become a major partner in building an interactive multimedia system in Japan. The system will combine NTT's digital network technologies, Silicon Graphics' digital media servers and object-oriented network software system, and the MIPS Technologies multimedia engine. "The interactive multimedia services system, in combination with the fiber optic digital network being introduced by NTT, will serve the expanding needs of a wide range of users," said President Masashi Kojima of NTT. (Note: Providing multimedia services is a key component of the national fiber optic network.)

Commscope's entry is due to the fact that its parent GI is a major supplier to Denver-based TCI, which started a cable television company venture in Japan with Sumitomo. The AT&T was also involved, providing its LaserLink package.

Planning the national fiber network has not been smooth sailing. For one thing Japan's Ministry of Posts and Telecommunications and NTT have different views on whether this network should be solely under NTT's charge or whether a number of carriers should participate. Again this increases tensions between the two Japanese bureaucracies.

Cost is also an issue: It is well into the hundreds of billions of dollars, a considerable sum even for the wealth that Japan enjoys. Budgetary concerns were magnified by the financial woes that recently afflicted the region. Early skirmishes in Japan's Diet were inconclusive, but there has not been wholesale approval of the nationwide project to date, although steps were taken to move forward particularly wiring major businesses with fiber.

As an early preparation for its national construction, Japan called for construction of culvert space to house the optical fiber cable to be installed throughout the nation. In all 150,000 km of duct space would be created for that purpose.

The idea of such a national construction is obviously very appealing to many fiber optic proponents around the world. A purely fiber network, able to offer any number of new services unencumbered by bandwidth restraints of any kind, stirs the imaginations of futurists. In fact constructing this national network, with the realistic incorporation of fiber-to-the-home and fiber-to-the-curb architectures, is the central mission of Full Services Access Network (FSAN), a global group in which the Japanese actively participate that is trying to reduce the cost of such systems.

The FSAN concluded early that an ATM-based passive optical network was a potentially useful solution, since such a network can be used with various associated architectures, including fiber to the curb, home, and cabinet, and it is not device-intensive. A key tenet of the working group is that the central fabric be easily allocated to a variety of end points. For example FSAN's Optical Access Networks group, which is designing the so-called engine for the new network, must avoid gilding the lilly, that is, designing systems that jeopardize commercial introduction due to cost, sources on the committee tell us. The group can go only as far as its members permit. If carriers decide they want a four-wheel car with air conditioning, for example, that is what the vendors will have to produce, one FSAN source told us, even though some vendors may also want to include bucket seats and automatic transmission.

While it makes sense to work together toward a consensus at some point vendors will want to break free to start selling the product. Carriers hope that will happen after enough uniformity has been introduced to avoid the SONET/SDH double standards, or even worse the VHS/Beta days of video recording. However all FSAN participants now seem to agree that uniformity is important, with a team emphasis on reducing costs so that the widespread replacement of nonfiber equipment can occur. In fact there is common agreement that the cost barrier is the nut FSAN is trying to crack. There is also agreement that FSAN costs will decline as production increases. The FSAN representatives further believe that at a sufficiently high-volume, development of new technologies is justified because it significantly reduces per line equipment and installation costs.

The FSAN in and of itself is not a standardization vehicle, it should be pointed out. Rather FSAN has attempted to incorporate applicable standards when they exist, and FSAN members include their specifications in appropriate standards. The FSAN hope is that developing requirement specifications will significantly advance deployment of cost-effective, full-service access networks.

The FSAN is divided into fiber-to-the-home and fiber-to-the-cabinet chapters. Both chapters are now involved in FSAN's fourth phase, which focuses on developing common business cases for deployment scenarios of interest to all members. Because it was established by telephone companies, FSAN does not address the needs of the cable television industry. It is not clear whether this situation will change when the two industries's networks are eventually integrated into one.

An important consideration continues to be interfaces between existing network architectures and FSAN products. The FSAN groups do not guarantee interoperability, but without FSAN there will be guaranteed inoperability, FSAN participants believe.

Despite the global effort the strong belief continues that the Japanese will develop national fiber to the home/fiber to the curb services before anyone else.

The United States is considered to have a variety of services, which argues against a homogeneous national fiber optics network. Even Bellcore's Paul Shumate, who spearheaded FTTC/FTTH efforts in the United States, believes there will be "a menu solution" in the United States, with no clear path only to fiber. Thus the national Japanese fiber initiative provides vendors there with a raison d'être to move forward with such products that vendors in other nations do not necessarily have (*FON*, 1 July 1996, p. 2).

Whatever happens, there is a race among vendors as costs begin to decrease. This will allow NTT and other carriers to present a more compelling case for such systems.

While Japan may not meet its deadline of 2005 for such a system, or even its backup deadline of 2010, fiber's value is too great not to be brought close to the home. The Japanese may very well be the first to do it.

The Japanese vision of an infocommunications society is the logical conclusion.

8

The Competition Down Under

How do we wire the sprawling, remote, yet progressive continent of Australia with fiber optics? The answer is competition, the first competitors: Telstra and Optus.

Telstra, Australia's established government monopoly, with a rich history of telecommunications achievement, says it has linked all of Australia's major business centers with fiber optics and the entire network is in the process of being completed. Wayne Asboth, who directs customer network planning at Telstra, said:

> Telstra made a decision some time ago to concentrate on deploying fiber as our backbone infrastructure, not only for the city loops, on which we provide SDH-type services, but also as part of the overall network infrastructure throughout the nation. (*FON*, 9 June 1997, p. 8).

Most Telstra domestic exchanges are interconnected by optical fiber, and Telstra also owns fiber linking Australia with the Australian state of Tasmania and New Zealand. Telstra offers businesses more than 120 products and services, including Freephone, virtual private networks, global frame relay, switched digital and satellite services. One example is Litestream, a high-bandwidth service that works much like SONET self-healing loops and adheres to International Telecommunications Union standards.

Telstra also uses fiber to deliver cable television, having provisioned 4 million homes with cable television and some 1 million with cable modems. However despite the enormous amount of money that both Telstra and Optus have spent on cable television, it has not been successful in Australia, perhaps due to the very good free television programs that are already available.

Telstra is also contemplating fiber connections to Papua New Guinea, and exploring alternative routes to the Asia Pacific region to handle its increasing role as a hubbing point for many services, including Internet. In fact the carrier is involved in an Internet telephony trial between Sydney and London, England. While it may seem unusual to have a trial that distance, a number of Australians continue to be only a generation or two removed from the United Kingdom, says James Shaw, a counselor at the Communications and Arts section of the Australian embassy in Washington, D.C.

Telstra had revenues of nearly $16 billion in 1997, with a pretax profit of $3.8 billion. The company had more than a million digital mobile customers in place and more than 84% of all access lines digitized by that time. Telstra intends to bolster its bandwidth internationally by deploying additional cable and introducing WDM technology. Alcatel is Telstra's international fiber and electronics supplier.

In 1998 Telstra announced that Pirelli Cables and Systems had been awarded an exclusive contract to provide Telstra with optical fiber and copper cable as part of an agreement through 30 June 2001, estimated to be worth from $123–$185 million Australian dollars. This contract is the result of Telstra's move to reduce the cost of network construction, which involved replacing MMCP, soon to be BICC Communications Australia, and Belden Australia, both of which had formerly also supplied Telstra, the carrier confirmed. As a result Pirelli increased its share of Telstra's domestic fiber optics business from 27% to 100%.

Telstra hopes to save some $62 million over the next 3 years as the result of various economies resulting from single sourcing. Pirelli promised to reduce inventory levels, lower logistics costs, provide much shorter lead times, and lower cable costs.

Telstra's John Losco announced that Telstra is carrying more international data traffic along its network than voice traffic. It is clear that the Internet has a global presence. Losco said:

> In 1996, our international Internet traffic was so low in comparison to voice that it scarcely registered.... You could see it start to lift off in early 1997, and then, in only 18 months, it rocketed up to exceed voice.

Losco believes that Telstra will carry five times the amount of data as voice by the start of the millennium (fibertoday.com, 7 May 1998, p. 2).

Cable & Wireless owns 49% of Australian service provider Optus, with the remainder owned by a consortium of companies, including Amp, National Mutual, Mayne Nickless, and AIDC. The Optus network is digital SDH. The company is also involved in Optus Vision, a cable television provider in Australia, and it is a mobile carrier. Optus began operations in November 1991, when it won a tender for a second carrier license.

The carrier laid its first cable on 5 March 1992, and by 1996 Optus completed its first full-year profit in the black by $60 million. By 31 January 1997, Optus had inaugurated a $90 million fiber optic cable between the cities of Adelaide and Perth; 2 months later it bought all of the equity in the Optus Vision project. The carrier deployed cutting-edge Fujitsu OC-192 equipment for its route between Sydney and Melbourne in 1998. That move also represents the first deployment of OC-192 in Australia, according to Optus, and it more than doubled capacity along the route. (Fujitsu had already provided the electronics for a 2300-km backbone loop between the cities, and it plans to expand that network north to Brisbane and west to Perth.) By 6 March 1998, Optus had opened a second optical fiber cable along the popular Sydney–Brisbane route.

In this first phase of competition, both Telstra and Optus were provided with very aggressive right-of-way provisioning, which led to aggressive cable installations and large towers that caused some discontent among the population. As a result laws were altered to provide Australian states with a more direct role in determining what is acceptable. Both carriers, indeed the entire continent, are benefiting from submarine cable trunks that come to Australia's shores. Both SEA-ME-WE 3 and Jasuraus provide the capacity.

With the exception of its cable television trial, which experienced some software problems that did not help overcome the general reluctance to pay for cable television services, Optus was successful: The carrier took 13% of the long-distance and international business by 1995 and increased that to 16% at the end of 1996.

Optus and other carriers had eroded Telstra's market share to 79% by 1997. The reality of liberalization is coming to the Australian marketplace. "The completion of this network expansion shows that Australia is capable of leading the world in the deployment of technology to support growing market needs," said Cliff McElveen, Optus group technical director.

Despite a few painful moments (Telstra staff reduction of about one-third and the less-than-spectacular results of the cable television trial) the first phase of Australia's competition was a success. There were approximately 46.50 telephones per hundred residents in 1997, costs stabilized and in some cases declined, and services increased. Based on those results the Australian government opened the entire nation to competition on 1 July 1997, with no bars to foreign ownership. "In 1997, real telecom competition took hold," says Shaw; "it is a duopoly no more."

Indeed due to liberalization twenty-two new carriers throughout the nation filed, offering a host of services. In a show of altruism Australia's Telecommunications Act of 1997 allowed those carriers to be as self-regulating as possible, although there is government backup if need be. As a result there are no restrictions on building or operating communications infrastructure; minimal restrictions on the type of technology used; no restrictions to entry to any telecommunications service market; and increased reliance on self-regulation, which is expected to result in a more efficient industry.

In preparation for full liberalization, Telstra spent 5 years upgrading its networks, systems, and products; claims improved customer service and increased efficiency and profitability. Telstra is philosophical about liberalization in its 1997 annual report:

> The Australian telecommunications market has undergone significant changes in recent years, with new players entering the market, and the prospect of more to follow.... While the introduction of competition has affected Telstra's market position, it has also stimulated overall market growth, and this trend is expected to continue.

One-third of Telstra is now privatized, and that is expected to increase to two-thirds. The only question is who is going to provide the universal service also called for in the act. One of the few negatives as telecommunications markets are deregulated around the globe is that carriers tend to migrate to where there are business opportunities. Those outside the profit centers have the potential of being excluded.

Australia, with a population of 18 million spread over a large land mass, is not immune to this problem, nor are other nations. In this situation government intervention is necessary to level the playing field. Yet weigh that challenge against the profound advantages of liberalization: Some 96.8% of Australian residences are in the process of obtaining ISDN capability. For its size and breadth the country has installed much fiber, a process that considerably hastened by competition. "New technologies are important to us in keeping up with the rest of the world," Shaw says.

Of course every issue has not been resolved. As in seemingly every deregulated nation, Australia is undergoing a tug of war over access fees with the entrenched carrier, who wants higher fees and the alternative carriers, who hope to lower these. There is also a public relations campaign at the National Office for the Information Economy to convince residents of the benefits of larger bandwidth and the services it can provide. "They try to pick the best examples around the world and enlighten people," Shaw says. This includes working with farmers to give them the latest information they require and also wiring schools and libraries with fiber to improve education and general awareness.

Opportunities continue to abound in Australia. The costly hybrid fiber coax constructions for cable television that did not show a profit may be excellent avenues for advanced Internet usage through cable modems. Since Telstra now provides a cable modem service, the nation appears destined to turn even its telecommunications mistakes into successes.

9

China Comes up Huge

If the twentieth century belongs to the United States, as some have suggested, the twenty-first century may very well belong to the Chinese (*FON*, 18 March 1996, p. 1). China clearly represents the most attractive market for telecommunications around the globe, one that will even surpass that of the United States within a decade, according to United States International Trade Organization, which acts as the Telecommunications Industry Association's representative in China.

The market has already attracted the attention of the major global heavyweights, as well as some local vendors. The list includes Alcatel, Siemens, Nortel, Ericsson, NEC, Fujitsu, ECI, and Hui Wei (local). One reason is China's increasing integration with the global economy, lowered trade barriers, growing international trade, and a strong influx of direct foreign investment, USITO says. China also planned to adjust its annual GDP growth target from 8 to 9% in 1998 to accelerate growth. (Note: The GDP was expected to range from \$5–\$6 trillion by the millennium.)

In the last years of the twentieth century and first years of the millennium, China is expected to inject \$750 billion into its infrastructure, with a large sum destined for telecommunications and fiber optics, according to USITO. Under the telecommunications system in place in 1998, China Telecom was the monopoly carrier and part of the regulatory body establishing standards! The USITO believes that China Telecom will be separated from governmental activities and eventually divided into several entities.

Currently China Telecom is one of the largest telephone companies in the world, with more than 100 million phone lines and some 8 million cellular subscribers. China Telecom has invested more than \$40 billion in a national information backbone. National ISDN and frame relay networks are under construction, and as of 1998 ISDN was offered in more than 20 cities. The ATM-based broadband networking is being developed in major cities.

China Unicom is China Telecom's central competitor, with a near-term goal of capturing 10% of the fixed-line market and 30% of the cellular market. Other players include Great Wall, which represents the Cell Division Multiple Access cellular arm of the government agency MPT, and cable TV, which dominates China's entertainment industry. There are currently four ISPs in China's rapidly expanding market. There were 1 million Internet users in China at the close of 1997, and that number continues to grow rapidly (see Figure 3.3).

If fiber optics has one inventor, it was Dr. Charles Kao, a native of China. The seminal paper written in 1966 was coauthored by Kao and G. A. Hockham, but it was Kao who actively pursued the phenomenon and remained active in promoting its benefits. Kao returned to his homeland in the 1980s to take a leadership position at the Chinese University in Hong Kong. Therefore it is not surprising that the Chinese have taken advantage of fiber optics.

The Chinese were assisted in their efforts by Lucent Technologies, which made a special effort to establish a market there. But the market is becoming increasingly open, and many vendors are trying to make inroads. In fact several reports suggest that China will be the largest national fiber optic market in the future.

The Lucent story began years ago when AT&T made a commitment to the Chinese marketplace that led to the installation of the first SDH networking equipment in China. As a result Lucent won over 300 SDH contracts in China, worth a total of approximately $650 million. The first Lucent SDH route between Hong Kong and Guangzhou was fully operational on 11 November 1994.

In one contract China's Ministry of Information Industry awarded Lucent projects totaling $62 million to extend China's backbone telecommunications network. As part of the contract, Lucent agreed to supply 2.5-Gbps transport equipment and a network manager system for the Guangdong Post and Telecommunications provincial long-distance transmission backbone and local area SDH networks. Lucent is also supplying China's Directorate General of Telecommunications with Lucent's latest transport and network management systems for the Shanghai–Jinhua–Naping–Guangzhou SDH backbone. This includes WDM equipment.

Lucent has seven regional offices in China, six joint ventures, and two wholly owned companies, with more than 2000 in-country employees. One of its projects, Lucent Technologies of Shanghai Ltd., is a joint venture that manufactures SDH systems. The plant was granted ISO 9002 certification in June 1995, making it the first telecommunications facility in China to receive such certification. According to Rau Chang, vice-president, Transmission, Lucent China:

> China's telecommunications infrastructure is increasingly becoming the most modern and state-of-the-art network in the world, particularly its transport network, with the latest SDH technologies.

Fiber has been a key to China's growth, and Lucent has been a leader in fiber. While there were approximately 8.7 telephone lines per 100 people in China in 1998, by 2020 China plans to have 40 telephone lines per 100 people. Considering the size of China, that is an extremely ambitious goal that translates into a "massive telecom infrastructure buildup," says Lucent's Mike Chan, who works in China. "Internet usage also is a major driver, as it demands more capacity from fiber optic networks," Chan notes. While there are now 1–2 million Internet users in China, Lucent estimates that will increase 10–20% each month.

According to Chan, Lucent believes the telecommunications liberalization now being undertaken by the Chinese government is improving its prospects. Chan told us in October 1998:

> To give you a couple of examples, Lucent Technologies recently announced that China's Ministry of Information Industry (MII) awarded Lucent multiple projects, totaling $62 million, to extend China's backbone telecommunications network.... And, a few weeks ago, we announced a $19 million contract for optical networking equipment with Fujian Posts and Telecommunications Administration in Beijing, which is expanding its telecom network to meet its customers' growing demand for voice and advanced services.

The bulk of China's population is in the eastern half of the country, which includes such major cities as Beijing, Shanghai, and Shenzhen. The Chinese economy is considered stable compared to other Asian countries that experienced problems in the late 1990s, Chan said. The entire Chinese market is still growing, although it is not growing so rapidly as it was in the past; inflation is under control; carriers have a very good reputation for paying their bills, Chan adds. Chan notes that each "boasts among the most modern networking infrastructures in the world." Lucent optical networking equipment represents the primary transport vehicle in each of these cities, Chan adds. And Lucent has not held back on its best technology elsewhere in China: In 1998 it announced the sale of digital television equipment to the Hubei Cable TV Network Company, Ltd., for what Lucent says will be the world's largest digital video network using fiber optic technology.

The sale complements earlier digital network contracts that Lucent Digital Video made in China, including the installation of the world's first optical MPEG-2/ATM/SDH network in Zhejiang, as well as a digital network in Hainan. The Hubei network is part of China's digital television infrastructure construction. Each of China's 31 provinces is expected to use MPEG-2/ATM/SDH over ATM as the standard platform for transmitting digital television. As part of the $4 million contract, Lucent will provide its award-winning digital video encoder and decoder, which are part of the Lucent Wavestar Digital Video System.

Particularly active in this effort was China Telecom, which played a critical role in constructing China–U.S., a planned $1.4 billion, 27,000-km network to connect the United States directly with China using two landing points apiece. When operational the link will directly connect China to the global community of fiber optics users and with an 80-Gbps trunk for major bandwidth capacity. Other carriers involved in constructing this network include AT&T, Hong Kong Telecom, NTT, KDD, Korea Tel, MCI, Sprint, SBC, and SingTel.

The consumption of one element of a fiber optic system, single-mode fiber, is expected to grow from $308 million in 1997 to $1.8 billion by the year 2007 in China, according to Electronicast Corp. The large majority of that growth will be in telecommunications, which accounts for $1.5 billion of the total.

China's most recent 5-year plan projects deployment of 65 million telephones and 96 million exchange lines by the millennium. The teledensity goal of the current 5-year plan, 1996–2000, is to reach 10–40% in the urban areas, Electronicast reports. By the end of the last 5-year plan, teledensity in China reached 3.6% for every 100 people, exceeding the goal by 1%. Most of the lines are fiber optic-based.

The Beijing–Jinan fiber optic line was the longest constructed in China by 1998. In 1995–96 a 2480-mile fiber optic cable linking Hangzhou, Fuzhounad, and Nanchang was completed. A network joining Beijing, Jiuyiang, and Guangzhou was also built. In 1997 China's MPT began constructing a 2754-m route linking northwestern China with Qinghai Province and Tibet, Electronicast reports.

The 5-year plan calls for eight east–west fiber optic links to intersect eight north–south routes through more than 60 major network nodes, comprising more than 100,000 km of optical fiber cable, says Electronicast. Much of the cable, which is planned to be deployed by MPT, will contain 36 fibers and operate at the 1550-nm wavelength.

Other U.S. vendors besides Lucent are finding the Chinese fiber optics marketplace to be lucrative. While noting that he did not want to overdramatize, Jim Granger, president of ADC Telecommunications's access platform systems division, observed, "It's big, sometimes it boggles my mind how big," referring to the potential market for hybrid-fiber coax networking (*FON*, 18 March 1996, p. 1).

Partners of ADC in its Chinese venture include China Unicom, the Ministry of Electronics Industry, and the China Communications Systems Corp. Granger believes that at least half of the 12 million lines that China Unicom will install will be HFC. This will involve ADC's Homeworx hybrid-fiber coax network.

The status of the Peoples Republic of China as a potentially lucrative market affected the U.S. government's policy on exports to China. It was clear several years ago that vendors in other nations were profiting from China's enormous SDH construction while U.S. vendors were held back by compliance with international trade restrictions. To rectify the situation the U.S. Department of State lifted all restrictions on SDH equipment and most fiber optics equipment; however there is still concern about direct sales to the Chinese military, as pointed out in a General Accounting Office (GAO) report:

> Broadband telecommunications equipment, such as ATM and SDH, has numerous civilian and military applications, and is becoming increasingly available in China as it strives to improve its telecommunications networks to meet

> international standards.... Since the removal of most export restrictions on telecommunications equipment, the market for such equipment in China has grown quickly and large quantities of SDH equipment have been sold to China to modernize its commercial long-distance network.... Advanced telecommunications equipment, particularly SDH, is increasingly used to be consistent with the emerging international communications standards, and it should vastly improve China's outdated and underdeveloped telecommunications systems. (*FON*, 9 December 1996, p. 3).

According to industry officials U.S. companies have sold tens of millions of dollars worth of SDH equipment in China beginning in 1994, and several joint ventures are currently manufacturing SDH equipment in China to build China's telecommunications networks, the report says. In addition, the report notes that the Chinese military is seeking SDH and ATM equipment, potentially to improve their command and control capabilities. Chinese military hospitals would also benefit from telemedicine from fiber optic deployment. The report suggests that more could be done to define characteristics of the Chinese marketplace so that U.S. vendors could continue to sell to the Chinese telecommunication market, but not support the efforts of the Chinese military.

Whatever the solution, it is unlikely to restrict U.S. companies from doing business in this robust new market, particularly as national economic security becomes more and more synonymous with national security. The Chinese commitment to the global information superhighway is simply too strong not to allow U.S. vendors to do everything possible to exploit it.

Section 4: Europe

Europe had much the same monolithic telecommunications structure on a country-by-country basis as did the United States and other nations in the 1970s and early 1980s. It is not difficult to understand why, since it made sense to have one telephone company responsible for the voice service that was implemented in these countries. And with no reason to go beyond telephones, and no extraordinary demand for that traffic, there was no reason for change.

However the monopoly often became intertwined with the government and certain suppliers. Then customers had no recourse but to pay whatever price the PTT wanted, perhaps with the government's compliance, for what had become an essential business and personal tool. Service often suffered because there was no competition and therefore no recourse for disgruntled subscribers.

A major step toward competition occurred in the United Kingdom in 1983, when Mercury, which had ties to the U.K. international carrier Cable & Wireless, was allowed to compete with entrenched carrier British Telecom (BT) due to complaints about the attitude of some BT employees, BT's promptness in responding to service calls, and so forth. Not surprisingly Mercury used fiber optics heavily in competing against BT. A central focus of most alternative carriers, was Mercury largely serviced the business community through its *figure of 8*, a fiber optics double-loop embracing a variety of U.K. business centers. By most accounts Mercury was an unbridled success, and it led the British government to open the market to more competitors at both the local and national levels. These included such carriers as Energis, which used the national utility grid to lay fiber, and City of London Telecommunications (COLT), a type of competitive access provider that concentrates on inking businesses to intracity fiber.

There were numerous ramifications of this U.K. liberalization throughout the continent. For one, it became clear that BT would have to do business outside the United Kingdom, if it were to keep revenues up, since there was no doubt that its domestic share was going through erosion. Yet BT also wanted to maintain and improve its domestic network to keep competitor gains as small as possible, and therein lies an ongoing dilemma for incumbent carriers.

In addition the United Kingdom benefited from foreign investment as carriers from the United States and elsewhere attempted to make a market in this new, open arena. Consumers were given more choices, and phone prices declined substantially.

The European Commission mandated that all PTTs in Europe be deregulated. The final bastion of the PTTs, voice communications, fell on 1 January 1998. (A few PTTs asked for and received extensions but only for a limited period of time.) The continent was being liberalized, albeit on a nation-by-nation basis. As in the United States, Europe had to be persuaded that an open telecommunication market resulted in more jobs, better service, lower rates, and a more vibrant economy. And this was not always an easy task, particularly in such nations as France and Germany, where France Telecom and Deutsche Telekom had played a dominating role for many decades. These governments and PTTs also correctly surmised that some employees were going to lose their jobs due to liberalization.

Yet in reality these monopolies had become fat and lazy to the detriment of their economies and customers. In many respects they had become too much like a government—not driven by profit and not worried about staying in business—with the resultant complacency. Outside vendors complained that France Telecom did not allow them to participate, since it had become closely allied with Alcatel and a few other vendors. Residents in Germany, whose monopoly Deutsche Telekom, had a similar relationships with Siemens, were unhappy about their monthly telephone bills, although the service quality was rated outstanding.

Nevertheless liberalization is occurring at both the national and continental level, and such carriers as WorldCom, Viatel, British Telecom, Global Crossing, and Hermes Railtel, are now laying optical fiber cable throughout Europe. A jubilant WorldCom completed the first phase of its pan-European construction on time, connecting the key cities of London, Paris, Amsterdam, and Brussels, then moving on also to connect Paris, Amsterdam, and Frankfurt.

Not surprisingly BT has reemerged in a variety of joint ventures to compete with entrenched European carriers. In fact BT is committed to a pan-European network that connects Belgium, France, Germany, Holland, Italy, Switzerland, and the United Kingdom.

Nortel Networks is helping BT in these efforts. It claims to be the premier STM-64 vendor in Europe as a result of the deal. British Telephone's European partners, in which it has a financial interest, include Albacom (Italy), BT Belgium, Cegetel (France), Sunrise (Switzerland), Telfort (Holland), and Viag Interkom (Germany).

While there were staff reductions in many of the major PTTs and their lead suppliers, there were also now ventures that employed at least as many people as were laid off and offered now services and opportunities to enlivening the European telecommunications market. In fact these new constructions led to a growing market throughout Europe for fiber optic equipment. The DWDM systems alone are expected to increase from a minuscule $37 million market in 1995 to $1.4 billion by 2000 and to $3.5 billion by the year 2005, according to Electronicast (see Figure 3.2).

Does that mean that Europe has caught up with the United States and Canada, at least in terms of providing advanced telecommunications services to its populations? By most estimates Europe still has some distance to go, although it is moving rapidly in the right direction. In particular the European community is still catching up with regard to Internet usage and providing access to bandwidth generally. The European Research Net, TEN-34, recognizes that its network operates at 45 Mbps while U.S. researchers connect at up to 622 Mbps.

An official from at least one European PTT believes his nation is opening up its market faster than is the United States to competition (and it is not the United Kingdom which besides Chile probably has the best case for making such claims).

PTT Nederland chief executive Wim Dik said that European telecommunications operators are not given the same opportunity to enter the U.S. market that U.S. telecommunications operators are given in Europe (*EuroInfo Tech*, 26 March 1998, p. 7).

The U.S. market is not so liberalized as it claims, and the European Commission should do more to extract fairer treatment for European operators entering the U.S. market, Dik said. Europe should be proud of creating the world's most open telecommunications market, worth $200 billion per year, and U.S. operators are present in many European markets, Dik observes.

Telecommunications deregulation was essential to building the global fiber optics superhighway. The best ideas from many quarters are necessary to operate this superhighway at maximum efficiency; such ideas grow and flourish best in an open environment. Every nation seems destined to learn this if it is a fully participating member of this network. It has in fact become a rite of passage to the globally fibered world of the new millennium.

10

Deregulation Shakes the Continent

Much fear and distrust accompanied telecommunications deregulation in Europe, largely fomented by affected PTTs and those governmental elements tied to them. In Germany for example Deutsche Telekom appealed to national concerns to keep a strong PTT in place—with some governmental sympathy. But prodded by the excellent and firm European Commission timetable, citizens in these nations saw the benefits of liberalization firsthand in greater choices for services and prices, and the economies of the countries involved were invigorated by new vendors and carriers.

Various vendors offering the abundant communications services now at our fingertips represents a natural evolution from voice-only services historically delivered by monopolies. While we recognize the wisdom and benefit from this evolution, try telling that to someone about to lose his/her job at one of the affected PTTs! Therefore a strong central body was required in Europe to mandate liberalization. The European Commission attempted and continues to achieve fairness and as equal a playing field as possible. There have also been calls for a continent wide organization to operate under that specific mandate.

In one respect nations with protected monopolies saw the economies and services of their liberalized neighbors thrive while their own economies continued to experience the problems of monopolistic organizations that often place their own needs above their citizens.

Anyone familiar with European liberalization knows that the pace of deregulation depended on the nations implementing it. Since the United Kingdom and the Nordic nations are substantially ahead of other European nations, they are better prepared to enter the new world of global telecommunication, to the benefit of their citizens. This was also the situation with some telecom services: Mobile communications operated in a liberalized telecommunications environment more openly than did wireline. The European Commission opened data services to competition far earlier than voice, which was finally liberalized on 1 January 1998.

An example of both trends is Ireland, where complete liberalization did not occur until 1 December 1998, when both voice telephony and the public infrastructure were liberalized at the expense of entrenched carrier Telecom Eireann. However elements of telecommunication services were opened to competition in Ireland far earlier; for example fax, data, and value-added services were liberalized in 1992, and 55 licensees now provide those services.

In Belgium Esprit Telecom was awarded the first independent telecommunications license. A public network operator (PTO) license enables Esprit Telecom to operate its own digital broadband telecommunications network across Belgium. Esprit acquired the raw fiber optic assets to build the network from Belgian railway operator Nationale maatschappij der Belgische Spoorwegen BV.

The forces unleashed by deregulation proved to be unstoppable. Monopolistic PTTs became leaner and more competitive, offering services they may not have been compelled to provide before and at more attractive prices. But more importantly capital was generated and new vendors entered the market, offering unique services and alternatives, providing the opportunity for people to rethink services and offer choices in satisfying communications needs.

The United Kingdom played a leading role in all of this. It wisely instituted a duopoly to overcome fears, then later upgraded this competitive framework to full competition. In this environment many trends of accelerating telecommunications deregulation quickly manifested themselves: Numerous national carriers enter the market, cable television systems increase, the price of phone calls decrease, and service often improves with alternative carriers serving the upper-business market. The monopoly continues to maintain its entrenched market, though market-share erodes, and the monopoly must often enter other markets to do battle with entrenched carriers there. Chaotic? Yes; beneficial to the nation and its citizens? Certainly; the best way to connect a nation to the global fiber optics superhighway? Absolutely.

In many respects once deregulation was in effect, the situation was analogous to other processes where a newer model replaces an older one, for example replacing typewriters with personal computers or eight-track tapes with cassettes. There is no looking back.

Fiber optics gives alternative carriers a fighting chance when competing with entrenched monopolies. Armed with optical-networking processes, the largest bandwidth available, unencumbered by the legacy PTT networks, alternative carriers often become a potent marketplace force in a relatively short period of time. Although the forces unleashed by liberalization may be temporarily slowed by such issues as high access charges, alternate carriers will find a way to compete and their markets will thrive.

Liberalization of the European telecommunications market is like a freight train that has left the station and is now picking up appreciable speed. It will proceed more rapidly, and it will deliver the products required.

11

The U.K. Testbed

When the U.K. decision was made in 1983 to allow limited telecommunications competition, it was considered in some quarters a radical departure. While most British citizens would agree that it was a worthwhile experience, it was a rather brutal process for British Telecom, which went from 246,000 employees in 1984 to 127,000 in 1997.

While it did not go so far as the AT&T divestiture in the United States, and there were other attempts at telecommunications deregulation elsewhere, the fact that a major European power with the visibility of the United Kingdom allowed another potent telecommunications force to compete against its entrenched telecommunications monopoly represented a critical breakthrough in how Europeans think about telecommunications. In years to come this decisive action in the United Kingdom may very well help define the nation as a telecommunications leader in Europe, just as it may help define the United States as a global and North American leader and Japan, China, and especially Australia as leaders in the Far East. In retrospect as the world community shares the benefits of global liberalization, the United Kingdom endeavor seems almost natural. But given the power that PTTs wielded in Europe, and indeed that entrenched telephone companies had around the world, it was in fact a courageous maneuver carried out against strong protestations from a powerful and entrenched monopoly.

The PTTs often coopted the ruling governments of the day, selected their own suppliers, in many respects set their own rates, and in general established a substantial power base. To many citizens, thanks to extensive PTT advertising campaigns, their success was identified as national success.

Against this backdrop the British government allowed competition in the form of Mercury. In permitting Mercury to compete against BT, the government also allowed Britain's only other major telecommunications power, Cable & Wireless, to take an 80% ownership. The alternative carrier was not then a token that Britain designated to provide weak competition so it could say that the U.K. market was being liberalized or a new entity that would be hopelessly low on the learning and capital curve. It was the United Kingdom's own international arm, which had successfully built a global network from grit and knowhow. BT had reason to be alarmed.

Cable & Wireless was also well-known to people and companies based in the United Kingdom, even though it initially used the name Mercury. This proved to be important later on. As with other alternative carriers that followed, Mercury did not attempt to build the full telecommunications infrastructure that the entrenched monopoly already had. That would have been too costly. What it did instead was build a new fiber optic infrastructure through some of the most lucrative business zones in the United Kingdom and offer large amounts of bandwidth at reduced prices to U.K. businesses.

While this was decried in some quarters as skimming cream or cherry picking, it provided a new carrier with the most financially beneficial avenue into the telecommunications market, thereby avoiding years of dismal losses before substantial revenues begin to accrue. And as often happens the alternative carrier responded to its customers networking needs by building networks throughout the United Kingdom.

When it became clear that the British economy was benefiting and that the alternative carrier provided essential services, the government decided to allow more widespread competition. This opened the cable television and the telecommunications markets, thereby placing the United Kingdom far ahead of most of its European neighbors.

There were a few twists along the way, Mercury continues to be a viable carrier force in the United Kingdom. Following a restructuring the company had approximately 8200 employees in 1996, and it reported continued growing call volumes and market share. In 1996 Mercury also sold a 75% share in its customer premises equipment business to Siemens and outsourced its national directory enquiries to U.K.-based Excell Agent Services.

Major investments also continued in SDH transmission and ATM switching. Maintaining its emphasis on corporate customers, Mercury launched private circuit services using SDH for Reuters and Midland Bank. Other U.K. business using Mercury's services include Great Universal Stores, the second largest U.K. retailer; the insurance company Norwich Union; and the U.K. National Health Service. As coowner of Gemini, a trans-Atlantic fiber optic network with WorldCom, Cable & Wireless built a national fiber infrastructure across the breadth of Britain to connect its submarine gateway with London, which tied into the Mercury network.

With the opportunity to participate in Mercury, the Australian Optus project and other alternative carriers, Cable & Wireless CEO Richard Brown reported in April 1998 that the carrier's revenues grew 15% in the prior 12 months, more than three times those of entrenched carriers BT, France Telecom, and Deutsche

Telekom. Brown observed that in a period of 12 months, Cable & Wireless increased its number of customers from 11 million to 16 million, with new customers joining at the rate of 50,000 weekly.

Recognizing the importance of staying ahead of the times, Cable & Wireless merged Mercury with three cable television companies to offer telephony, information, and entertainment on the same network. The result, according to Brown, "was a step-change in revenues, in margins and the critical mass to start doing new things. Now, our U.K. customers are getting a taste of integrated services."

The number of homes in the United Kingdom with Internet access through a personal computer increased from 100,000 in 1994 to 1.6 million in 1998, Brown said, noting that Cable & Wireless is already "the world's most international communications company," operating in 72 nations. "We own one-twelfth of the world's international fiber optic capacity and 40% of world capacity for installing and maintaining undersea cables," Brown said.

Another growing U.K. telecommunications carrier is Energis, PLC, which focuses exclusively on the business sector. The carrier entered during the second wave of U.K. telecommunications liberalization and now has more than 5000 km of optical fiber cable installed throughout England and Wales along the National Grid's electricity pylon infrastructure. The carrier also has a capacity-sharing agreement with Scottish Telecom, as well as an international license. Like other long-distance carriers, Energis recognizes the importance of having access to the local exchange; however that can be an expensive proposition, even though some of Energis's network encircles and runs through some metropolitan areas.

Still to give it further clout, Energis announced in 1998 the launch of Metro-Holdings Ltd., a joint venture geared to the metropolitan market and undertaken with Deutsche Telekom and France Telecom. It is concentrating on such major cities as London, Manchester, and Birmingham. The company is 50% owned by Energis and 25% owned by each of the PTTs. The new constructions will use state-of-the-art SDH equipment along the new fiber optic network, the same as that included on Energis's national construction, the carrier says.

Later in 1998 Energis announced that it had acquired Planet Online Ltd., the largest independent U.K. Internet service provider focused on the business sector. Planet is now a wholly owned subsidiary of Energis. The cost of the acquisition was $120 million. Energis says it carries approximately 40% of U.K. national Internet access traffic, and it is enhancing its position in this market by deploying recently developed carrier-scale IP technology.

By acquiring Planet Online, Energis augments its portfolio value-added Internet and Intranet services and a substantial corporate customer base. The combination extends Energis's capabilities and range of data business services. It provides cross-selling opportunities to both organizations and customers and adds a critical mass of Internet sales and technical staff.

Planet Online was founded in July 1995 and employs more than 150 people. Customers include Airtours, Barclays PLC, Bradford and Bingley Building Society, Cadbury, Midland Bank plc, Mirror Group Newspapers, the National Lottery, PowerGen, Prudential, Reuters, Sony Computer Entertainments, Thomas Cook, Yorkshire Electricity, and Yorkshire Water.

Yet another upcoming U.K. carrier is COLT Telecom Group PLC, which began as a small fiber optics carrier in the City of London business district, and then offered fiber optic capabilities in 12 key European cities by mid-1998, with plans to service 26 cities by the end of the year 2000. The COLT is growing so rapidly that its revenues for the first 6 months of 1998 exceeded all of its revenues for 1997. The company is not profitable, but it does anticipate that it will begin making money by the year 2002.

Fiber optic networks that were already operational by mid-1998 include London, Paris, Frankfurt, Hamburg, Munich, Berlin, Zurich, and Amsterdam. (fibertoday.com, 17 July 1998, p. 1). The carrier had 728 fiber route km operating, a 123% increase over figures for the previous year. It also had 1591 buildings connected by mid-1998. One substantial customer is Racal Telecom.

The COLT has announced that it acquired Telecom Noord West NV for approximately $568 million. The Dutch carrier has installed about 1000 route km of fiber networking and associated switching and transmission equipment in the Netherlands, including 170 route km serving key Amsterdam business districts.

Such growth was not limited to carriers operating in the United Kingdom. Bookham Technology, a company specializing in optical devices using a unique chip-making process known as ASOC, offers a world of silicon-integrated optics to customers; such optical devices include new DWDM components. Based in Oxfordshire, Bookham experienced a rapid increase in investments in the mid-to-late 1990s and went public in April 2000. Kymata, based in Scotland, is another strong new vendor.

Very few in the United Kingdom would dispute that there was a certain amount of chaos as the national communications infrastructure was liberalized. Likewise very few would dispute that their options and capabilities (to talk to each other, communicate by fax, use the Internet, and be entertained) were not substantially improved by the liberalization process.

In 1998 the United Kingdom was far ahead of many of its continental neighbors when liberalization took effect throughout Europe. Rates had decreased, its longtime monopoly adjusted and entering other markets, fiber optics was being used throughout the country. The United Kingdom had become a hub along the global fiber optics superhighway.

12

Deutsche Telekom: Fibering the East, Fighting Competition

With the breakup of the Soviet Union and the general demise of communism in Eastern Europe came many tremendous challenges. For Deutsche Telekom, the largest carrier in Europe, that challenge included rebuilding eastern Germany's telecommunications structure, which had been neglected under communist control. Deutsche Telekom took this mission extremely seriously, considering it an opportunity to reunite the nation into a strong telecommunications infrastructure. Its tool of choice for modernizing the East German network: Fiber optics.

The network, which was completed in 1997, is considered "the most modern telecom infrastructure in the world," Deutsche Telekom spokesman Hans Ehnert observed. Much of the network provides fiber-to-the-home capability, and new fiber throughout the former East Germany. In fact the East German network is now considered more modernized than West Germany. "We have 1.2 million homes in East Germany wired with fiber," Deutsche Telekom Chairman Ron Sommer noted. "In some respects, it is easier to start from scratch in building a new network" (*FON*, 10 February 1997, p. 1).

From 1988–93, Deutsche Telekom undertook the most advanced fiber-to-the-home architectures at that time, using optical access lines (OPAL), which consisted of seven close-in fiber projects. Its ramifications are still felt throughout the world. When Deutsche Telekom began the national telecommunication reunification process, there was very little to build on. "There was nothing left," recalls Ehnert, "and what was left, was very, very old-fashioned."

The entire East German effort included cobbling together 3 million homes with new switches, new cable lines, and new telephones. The project had a total cost of $26 billion. East Germany, as well as West Germany, is also benefitting from Netscape-supported Internet services as the result of the fiber build.

What did Deutsche Telekom obtain for this seemingly utopian act? The answer is what Sommer calls the "toughest competition in the world" as deregulation settles into Germany, a process that is leading Deutsche Telekom to reduce its work force from 230,000 to 170,000 by the millennium. Yet Sommer, in an International Press Colloquium in early 1997, observed that Deutsche Telekom is making progress:

> Within the space of 7 years, our management and staff have transformed what used to be Deutsche Bundespost Telekom, an administrative ministerial authority, into a global player in telecommunications.

Sommer was personally involved with upgrading marketing and sales efforts, something he acknowledges Deutsche Telekom was not very good at in the past.

To fortify itself the carrier has roughly 150,000 km of fiber installed throughout its network, according to Ehnert. Deutsche Telekom embarked on an aggressive strategy to use digital subscriber line, a copper-based technology, for last-mile applications. However using these broadband trunks to the home will necessitate bringing fiber closer as well, from the switching nodes to xDSL hook-ups. There should be no fears that Deutsche Telekom's R&D arm will retreat as competitive pressure intensifies. In 1995, Deutsche Telekom spent approximately $816 million on R&D, and it anticipates spending about $1.2 billion in the year 2000, Sommer said (*FON*, 12 February 1996, p. 5).

In its successful initial public offering Deutsche Telekom sold 26% of its shares; shareholders include 156,000 of its employees. Deutsche Telekom also has a strong West German ATM service connecting to SDH, which potentially provides videoconferencing, multimedia, PBX, and LAN interconnections, in addition to widespread ISDN offerings for residential use. According to Herbert May, a member of Deutsche Telekom's board of management:

> In connection with fiber optic networks, ATM transmission technology is the foundation for the multimedia telecommunications of the future....
>
> ATM makes it possible for the first time to send large volumes of information of every kind simultaneously via a single access: telephone, fax, data or videoconference calls can all be made via an ATM access, despite the differing transmission bandwidths required.

The carrier's T-Net ATM service seems a natural for business. "The flexible network will allow access lines, giving capability from 2 Mbps to 155 Mbps," Georg Kronseder, Deutsche Telekom's product manager for ATM and DATEX M products, observed. Key customers are expected to include insurance companies, banks and the automotive industry. Auto manufacturer Porsche is a customer. Gerd Tenzer, also a member of Deutsche Telekom's board of management, goes one step further:

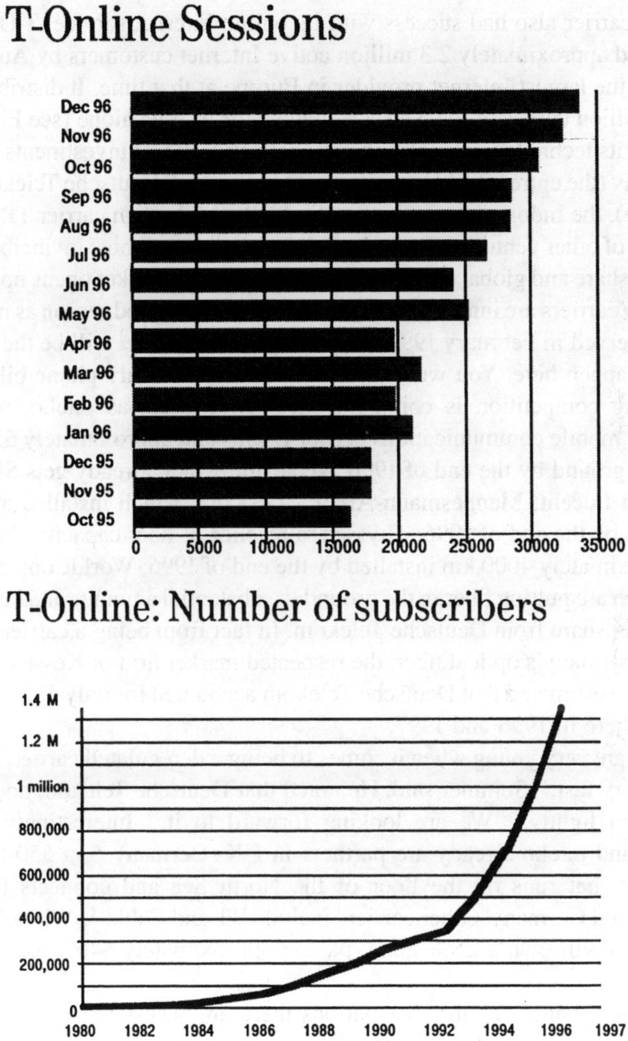

Figure 12.1. Deutsche Telekom has seen major growth in its Internet service, T-Online. (Photo courtesy of Deutsche Telekom)

> By expanding the world's most capable optical fiber infrastructure and using ATM transmission technology on a nationwide basis, Deutsche Telekom has made Germany a world leader in telecommunications.

The carrier also had success with its online Internet service, T-Online. The service had approximately 2.3 million active Internet customers by August 1998, making it the largest Internet provider in Europe at that time. It distributed more than 10 million copies of the Netscape Navigator in 1996 alone (see Figure 12.1).

With its technologically strong national network and investments in Global-One, Matav (the entrenched Hungarian carrier in which Deutsche Telekom owns a 67% share), the Indonesian carrier Satelindo, the Malaysian carrier TRI, Energis, and a host of other ventures, Deutsche Telekom appears almost invincible in terms of market share and global connectivity as the German market opens up. However alternative carriers are interested in this market, and for good reason as my taxi cab driver observed in February 1997, "Opening up the market will be the best thing that can happen here. You won't believe what my monthly phone bill is!"

Strong competition is coming from such carriers as o.tel.o, which also purchased mobile communications carrier E-Plus and approximately 6500 sheath km in the ground by the end of 1996; Vebacom, which already gets SDH equipment from Lucent; Mannesmann-Arcor's DBKom, which installed about 5000 sheath km by the end of 1996; Thyssen, a reseller of fiber capacity; Viag, which had approximately 4000 km installed by the end of 1996; WorldCom; and COLT, all of which are putting fiber in the ground in substantial amounts and beginning to take market share from Deutsche Telekom. In fact from being a carrier providing 100% of Germany's optical fiber, the respected market firm of Kessler Marketing Intelligence estimated that Deutsche Telekom accounted for only 50% of the fiber installed there in 1996 and 1997.

"We are very young when it comes to being a deregulated carrier, but we are moving very fast," Sommer said. He noted that Deutsche Telekom does not take competition lightly: "We are looking forward to it." Interestingly Deutsche Telekom and o.tel.o already are partners in UK–Germany 6, a 550-km optical fiber cable that runs on the floor of the North Sea and connects the United Kingdom and Germany. Other carriers include BT and Cable & Wireless. Each of the partners will own a 25% share, paying approximately $46.5 million for the privilege.

Not surprisingly as in other nations there are access issues, and Deutsche Telekom is using every argument to stave off competitors, but competitors, and the national telecommunications network will continue to thrive, compete, and grow, becoming a German leader as Europe opens up its telecommunications markets once and for all.

13

France Adjusts—At Times Painfully

It is probably fair to say that none of the PTTs welcomed telecommunications liberalization: It was a shock to their corporate and cultural systems, and they knew they would lose business. Perhaps no PTT was dragged kicking and screaming into telecommunications liberalization so much as France Telecom, which had consolidated operating revenues of $27 billion in 1997 and employed 156,620. Labor unions, which are strong in France, decried the potential loss of jobs to the PTT, even though such job losses are part of any liberalization process.

But perhaps the most painful alteration due to liberalization was the partnership between the fourth largest carrier in the world and its central supplier, Alcatel. Alcatel had built huge copper-producing factories, largely to outfit France Telecom, which had served the carrier for years. The two had contracts beneficial to Alcatel, which in many respects led to a bloated workforce that would never survive in a world governed by genuine global economics. At trade shows where numerous presentations described what new peaks vendors had reached with fiber, Alcatel always presented a copper-oriented paper. And since France was detached from other markets, it allowed these conditions to exist and even flourish. As a result France Telecom was saddled with what was becoming an outmoded technology.

France Telecom gradually recognized that its neighbors' more advanced networks deployed fiber optics communications systems, and that it would confront future competitors with generationally improved networks. And while recognition of the importance of bandwidth may not have been quite so important in the mid-1990s as today, there was a growing awareness that an abundance of clean bandwidth was desirable, as were a variety of vendors.

As a result Alcatel and France Telecom had to part company, which led to Alcatel closing 20–30 copper facilities in France, Germany, and Canada. Alcatel lost the staggering sum of $811.4 million in 1995 alone, and it had to endure continued layoffs in 1996 and 1997. Reason cited for these losses was privatization, in effect the loss of Alcatel's special relationship with France Telecom as a government monopoly and the ensuing lower market prices for cable and diminished interest in copper cable. "Optical fiber, which is being used more and more in Europe, has brought more competition to this market," conceded Alcatel Cable spokesman Michel Jamard in a classic understatement (*FON*, 29 January 1996, p. 1).

Not surprisingly Alcatel has since turned to fiber optics to become a leading supplier on the continent, dedicated to optical fiber production levels of 6 million km annually by the end of 1997. New facilities were constructed in Douvrin, France, and Claremont, North Carolina. Touring the Claremont facility on 8 August 1996, it was impossible not to be impressed by the enthusiasm of the workforce, the quality of the equipment, and the factory's commitment to producing optical fiber and cable. The fiber and cable plants, which are located side by side, employed 720 at that time. "Over the next 2–3 years, we will be spending a lot of money here," said C. J. Phillips, president of the facility. Both were then running 24 hours a day, 7 days a week.

Perhaps Alcatel's greatest success in fiber optics came as the leading contractor for the submarine fiber optic cable market. For years Alcatel and AT&T Submarine Systems, Inc. (SSI) vied for maritime fiber jobs. The SSI remains a strong competitor now as part of Tyco, but it became more involved in the Global Crossing venture. Bill Carter, who left SSI shortly after it was purchased by Tyco, is a principal in Global Crossing. This lead role in contracting opened many doors to Alcatel's cable and transmission systems, which are often selected when Alcatel obtains the job as prime contractor.

With regard to fiber optic transmission systems, one of Alcatel's earliest successes came in the area of digital cross connects, which are essential in connecting fiber optics to edge networking that is generally not fiber optics-oriented. Alcatel and Tellabs share that market globally, and it has been quite lucrative for both. In one 6 week period Alcatel sold $82 million worth of digital cross connects to MCI (*FON*, 3 March 1997, p. 4).

Alcatel was also making progress in the fiber optics transmission equipment sector. For example the vendor reached a strong customer base in the short time it was in DWDM. Customers include IXC, LCI, Entergy, WorldCom, AT&T, ACI, Iowa Network Services, TU Electric, Cable and Wireless, and Global Crossing. In fact Iowa Network Services selected Alcatel as its exclusive optical networks provider as part of a 5-year contract, including SONET equipment and Alcatel's 1680 Optical Gateway products, in addition to its optical add–drop multiplexer, the 1640 OADM. Alcatel estimated 1998 revenues from DWDM systems at $120 million; not bad for the company's third year of involvement and a testament to the explosive growth now occurring in the industry.

The WorldCom contract is of particular interest, since WorldCom was an important customer of Ciena's in the United States. When WorldCom announced a major European construction, Alcatel was able to obtain the contract, dealing a stinging blow to Ciena and establishing a major Alcatel preeminence on the continent. A critical advantage of large optics vendors like Alcatel, Nortel, and Lucent over smaller vendors, such as Ciena, is their capability to provide turnkey optics. In the case of Alcatel's European construction for WorldCom, the system includes SDH, optical amplification, and DWDM.

In the case of Entergy, a global energy company based in New Orleans with one of the largest private telecommunications networks in the United States, obtaining the contract showed that Alcatel's DWDM products have clearly entered the U.S. market. It also showed that Alcatel's 16- and 40-channel equipment was successful in the field.

Any DWDM systems contract for Alcatel is also a victory for its DWDM components group. In the case of the Entergy contract, Alcatel's DWDM transmitters are being used on the Alcatel 1648 SM, which eliminates the need for transponders on SONET channels. Alcatel believes in interoperating DWDM equipment and SONET equipment, not in eliminating SONET from the network.

Alcatel also made inroads with SONET and other optical equipment, signing a contract whereby it became a strategic fiber optics electronics supplier to AT&T. It was also part of a multimillion dollar deal with Canadian-based Fundy Telecom to supply maritime SONET rings, and it is supplying SONET equipment to IXC Communications and New Brunswick Telephone.

To build up its transmission equipment side and to expand its market share in the United States, particularly among the RBOCs, Alcatel purchased U.S.-based DSC Communications. In doing so Alcatel took a competitor and turned it into an important ally. With completion of the DSC purchase, Alcatel America, Inc., in fact now accounts for 21% of Alcatel's market.

There is no question that Alcatel has placed huge resources in its fiber optics efforts, shifting some personnel from its heavily overblown copper base. For example it operates a 700-employee R&D facility in Richardson, Texas, a portion of which is involved in photonics networking.

Alcatel is also reaching out in Europe, where it signed a contract to provide SDH equipment to the Unisource partners, which include Telia of Sweden, PTT Telecom of the Netherlands, Swiss Telecom PTT, and Telefonica of Spain.

In a move demonstrating how completely it has adjusted, Alcatel is now doing business with a chief France Telecom competitor, Telecom Development, a joint venture between French energy company Cegetel and SNCF, the French state rail company, worth $171 million (*FON*, 21 July 1997, p. 8).

Alcatel's gains were achieved despite some early missteps. For example Alcatel was hurt by its claims about fluorine-based optical amplifiers, which never had any market impact. Fellow scientists were critical of fluorine from the outset, saying it was not stable enough a base for optical products. In responding Alcatel seemed stubborn and defensive.

If Alcatel adjusted, so too must France Telecom, although supported by labor union unrest, it was able to delay implementing liberalization. To render the transition as painless as possible, France Telecom instituted a rebate-rebalancing program to fend off competitors. From the beginning of 1996 to the end of 1997, the carrier reported that its cumulative rate cuts for domestic and long-distance and international calls amounted to 47 and 48%, respectively. The carrier also became involved with a variety of international ventures, such as GlobalOne with Sprint and Deutsche Telekom; with Energis and Deutsche Telekom in the United Kingdom; and with TelMex in Latin America.

Besides Telekom Development, other competitors in France include SNCF and Cegetel, which combined have some 8000 km of optical fiber cable installed, and they are planning more; a new venture, NETs, which is building a London–Paris network; WorldCom; COLT; and Hermes Railtel. No doubt there will be more as this market opens.

Although France opened the door, hopefully it will support full competition and more aggressively construct its fiber optic infrastructure. In the meantime other nations scurry ahead, fueled by the power of innovation and a free enterprise system in which the government no longer plays a controlling role on the side of the PTT.

Alcatel continues to produce fiber as a licensee of the AT&T–Corning cross-licensing agreement, Alcatel officials told us. The agreement also applies to Alcatel's optical-fiber-making facility in Roanoke, Virginia, which Alcatel purchased earlier from ITT with the license. The Alcatel Claremont facility is ISO-9002 approved, and the cable facility is ISO approved.

Section 5: South America

With the possible exception of North Korea, no nation wants to be excluded from the global fiber optics superhighway, and this is certainly true of the South American nations, although they follow a variety of blueprints.

The deployment of optical fiber in Latin America is expected to grow more than fivefold from 1995, when it was less than 1 million fiber km, to the year 2002, when it is expected to be nearly 6 million fiber km, according to Pioneer Consulting. Pioneer believes the market for optical fiber cable will increase from $307 million in 1997 to $1.3 billion in 2002:

> Equipment manufacturers are doing better in this region than international telecom service providers.... Stringent import restrictions have been lifted in many major markets, including Brazil, which has allowed major vendors such as Cisco, Bay Networks, 3Com, DEC and Hewlett-Packard to begin selling equipment through local distributors.

Consistent with European nations some South American governments are more willing to let wireless providers enter than wireline carriers.

Following the trend of other nations, South American fiber will move from the interexchange market to metropolitan areas, according to a report published by Kessler Marketing Intelligence (KMI) entitled "Single-Mode Fiber-Optic Cable Markets in Latin America, 1997–2003."

Long-distance installations will lose market share in 2003, compared to what they had in 1997. The largest market will be metro/access, which will increase from 29% in 1997 to 44% in 2003, the report says. Cable television also represents a thriving market, expanding from 10% of the South American market in 1997 to 13% in 2003.

Fiber deployment in individual nations varies widely, due to differences in geographic size, economic growth rates, business expansion, and investment by operators. The largest markets are Brazil and Argentina; the smallest markets are Ecuador and Paraguay. Telecommunications liberalization was the single most important factor in the growth of fiber optics in South America, KMI says. The strong economic performance of Latin America in general also contributed to the growth of fiber optic systems. "Prosperity has driven demand for new services, brought money to fund network expansions, and generated customers for new services," the report says.

14

A Continent Demanding to Keep up

South America will not be excluded from the global fiber optics superhighway. Fiber is being installed throughout the seas and on land, driven by a generally thriving economy and telecommunications liberalization. The overall long-distance market in Latin America in general is valued at $50 billion in 1998, and it continues to grow at a brisk rate.

Electronicast estimates that the South American fiber optics component market will expand from $252 million in 1997 to $1.5 billion by 2007. The average annual growth from 1997–2002 is 18.6%, which is expected to increase to 20.9% from 2002–7.

Yet in many respects much still needs to be liberalized, and that will take some time to occur. The KMI analyst Kurt Ruderman notes that there is caution and the long-distance market is not completely open to competition. He characterizes it as being similar to the United States at about the time of the AT&T divestiture or perhaps even a little before.

The leader by far in terms of sheer size is Brazil, whose state-run telephone company, Telebras, is already privatized. A variety of domestic fiber optic deployments are underway in Brazil, including a coastal festoon network of submarine cables, as well as terrestrial systems running along the railway system, Pioneer reports. Long-distance carrier Embratel is investing $942.7 million through the year 2001 to develop its fiber optic network, Pioneer reports. Phase 1 includes expenditures of $595 million and covers some 10,451 km of fiber connecting the main cities.

KMI expects the Brazilian market to expand its optical fiber cable deployment from 2.3 million fiber km in 1998 to 9.5 million fiber km in 2003. In fact Brazil's Embratel is the only carrier at the time of the report that KMI expects to use WDM equipment. Brazil accounts for one-third of the total South American economy.

A main spur for Brazil was privatization of Telebras, which KMI reports

> Will push the Brazilian fiber market steadily ahead for the remainder of the forecast period as new owners position themselves to compete against one another in the near future.

The U.S.-based Reltec is bringing extensive fiber expertise to Brazil as part of a joint venture with Splice do Brasil Telecommunicacoes e Eletronica S.A. the new company, Reltec Sistemas de Energia Ltda., will also try to service other markets from its São Paulo headquarters.

Telefonica has a presence in Brazil, since it joined with consortium members Participacoes Rede Brasil Sul, Citicorp, and TI affiliates Telefonica de Argentine and CTC of Chile to gain 35% of Companhia Riograndense de Telecomunicacoes, which operates in a southern Brazilian province. Pirelli, Alcatel, and Furukawa have established vendor footholds in Brazil, Ruderman says. MCI is also active here, purchasing Embratel as part of privatization, as well as a one-third interest in Proceda Technology, a Brazilian service provider.

Next in overall deployment is Argentina, which is expected to grow in fiber km from 755,000 fiber km in 1998 to 1,982,000 fiber km by 2003 and which is also undergoing privatization. However the two primary carriers, Telefonica de Argentina and Telecom, could maintain rights through the year 2000, Pioneer Consulting reports. KMI's Ruderman characterizes Argentina as largely a dipole that has begun to open up. Other local carriers in the São Paulo area serve the wealthy areas there. Long-distance markets are expected to be deregulated first, followed by local access in Argentina.

Telefonica de Argentina made a commitment to digitize its network completely at the cost of $800 million. It is also part owner in Cablevision, a cable television operation in Argentina. Siemens and Pirelli made successful inroads into the Argentinean fiber market, says KMI's Ruderman.

Argentina is followed in fiber activity by the wildly entrepreneurial culture of Chile, which was the first Latin American nation to reform its telecommunications sector. The Chilean market is expected to grow to an installed base of 1.9 million fiber km in 2003, a substantial increase from 676,000 fiber km in 1998.

The four companies that provide the bulk of Chile's telecommunications requirements include Compania de Telecomunicaciones de Chile (CTC), Entel, VTR, and Telex Chile, according to KMI. Their operating units provide a full menu of local, long-distance, wireless, cable television, and other services.

Chile is one of if not the most deregulated telecommunications nation; it has ten licensed long-distance operators, eight of which were in operation in 1997. The four largest account for 96% of the traffic. Entel, the largest, has 42% of the domestic long-haul market and 33% of the international market. Chile in fact has three main backbones, including CTC, Entel—which is partly owned by Telecom Italia, and Chilesat, which has had some economic problems, Ruderman says.

The CTC is the largest local exchange provider, KMI reports, with a 90% marketshare as of the end of 1997. It is now owned by Telefonica de España (43.6%), foreign funds, employees, and others. From 1987–91 the telecommunications sector in Chile grew at an annual rate of 19.4%, reports Pioneer. The largest non-CTC provider is Telesat. CTC Mundo, a CTC subsidiary, is the second largest long-distance carrier in the nation.

According to Ruderman liberalization at one point veered out of control, with companies building fiber and telecommunications infrastructure seemingly everywhere. One sign of restored stability is that carriers now work together, trading fiber when it is practical.

Chile is followed by Colombia, which is expected to grow to a cumulative base of 921,000 fiber km by 2003 from 351,000 fiber km installed by the end of 1998. Colombia is being driven by competition among operators who are moving into each other's operating regions. However telecommunications companies in Colombia are all state run or run by municipal markets, says KMI's Ruderman: "The unions are very strong and have prevented the municipalities from privatizing." As a result joint risk ventures were established where such vendors as Nortel, NEC, and Siemens invest, then share profits for 7 years; after that the venture is turned over to the municipality. Competition in Colombia is due to long-distance carriers now beginning to compete in cities with the local carriers. The national utility power grid, ISA, is building a fiber optics network to compete with Telecom Colombia. Areas that require only basic telephone services continue to receive copper networks; however more affluent business areas normally receive fiber.

Next in market size is Peru, also an early leader among South American nations in implementing reform. Peru is expected to liberalize voice telephone services in 1999 in long-distance segments. Peru is expected to attain an installed base of 730,000 fiber km in 2003, an increase from 336,000 fiber km in 1998, KMI reports.

While some have questioned BellSouth's aggressiveness in the United States in forming partnerships with other carriers, it is moving rapidly in some South American markets, including Peru, where its Tele2000 network is challenging Telefonica de Peru, particularly in Lima. BellSouth, a fiber optics innovator in the United States, is also moving forward with networking in Brazil and Argentina.

Telefonica holds a 31.5% stake in the Peruvian operator, which exclusively provisions local, national, and international long-distance services, data transmission, cellular, paging, cable multimedia television distribution, pay phones, yellow pages, and value-added services. Peru has a government-instituted program in which each of the carriers pays a percentage to help fund networking in poorer rural areas.

The KMI projects Venezuela will reach an installed base of 680,000 fiber km in 2003, an increase from 217,000 fiber km in 1998. Venezuela is building a domestic shoreline submarine system to allow the Americas-1 fiber optics network to land on its shores, and it is embracing competition and foreign investment in telecommunications, reports Pioneer.

When Venezuela privatized its telecommunications operators in 1991, more than $1 billion was rapidly invested in the nation's infrastructure, which represented an enormous increase over prior years' spending, Pioneer says.

At World Trade Organization (WTO) meetings, Venezuela committed to opening markets for facilities-based voice telephone services in all market sectors as of November 2000. Full competition in facilities-based telecommunications services is anticipated without a phase-in period.

The GTE is prevalent in Venezuela as both the local and long-distance wireline carrier. One U.S. vendor to make inroads in the Venezuelan market is Advanced Fibre Communications, which was awarded a major contract by CANTV, Venezuela's national telephone company. According to Magaly Rangel de Pena, CANTV's director of technology planning:

> The UMC 1000 will allow us to deploy fiber-to-the-curb and fiber-to-the-building solutions to quickly and effectively satisfy telecommunications needs throughout Venezuela.

The KMI reports that Bolivia will reach an installed base of 179,000 fiber km in 2003, an increase from 77,000 fiber km in 1998. At the WTO meetings Bolivia also offered to phase in competition in all domestic and international services by November 2001, Pioneer reports. As part of the effort local voice telephony is to be provided by 16 exclusive local suppliers. Bolivia is offering full competition without a phase-in period for all basic services.

The KMI sees Uruguay increasing to an installed base of 130,000 fiber km in 2003, an increase from 57,000 fiber km in 1998; Ecuador expanding to an installed base of 100,000 fiber km in 2003, an increase from 27,000 fiber km in 1998. Paraguay, KMI reports, will increase its installed base to 45,000 fiber km in 2003, growing, an increase from 18,000 fiber km in 1998.

Since Ecuador postponed privatization, it is behind in fiber installations, Ruderman notes. It has state-run systems operating in the northern and southern parts of the nation.

There are already attempts to serve various nations in South America. One such venture seeking a broader footprint is Telefonica–Pan-American–MCI (TPAM), the joint venture formed by Spain's international telecommunications arm and MCI specifically geared to the South American marketplace. The venture is 51% owned by Telefonica and 49% owned by MCI.

The TPAM intends to build an all-digital network using fiber optic cables to link major business centers throughout South America. The intent is to carry large amounts of voice and data, according to MCI. By the year 2001 the network is expected to connect nearly a dozen business centers in Latin America, with gateway connections to MCI WorldCom and Telefonica facilities in North America, Europe, and throughout the world.

Probably the most exciting ventures are the host of submarine networks that are tying this continent together: The Pan American Cable system, Maya-1, Americas-1, Atlantis 2, the Pacific Transit Cable, Pan-American Crossing, and potentially Project Oxygen (see Figure 3.4). Interior South American carriers are already building and interconnecting with seaside South American carriers that will interconnect with some of these large undersea networks to obtain submarine cable exposure, says Ruderman.

While AT&T was an early leader in the Pan-American Cable system, MCI also emerged as a premier force in concert with Latin American carriers, Pioneer reports. The cable is connecting the U.S. Virgin Islands with Aruba, Venezuela, Colombia, Panama, Ecuador, Peru, and Chile. The six leading investors are AT&T, Telefonica de Peru, MCI, Telecom Italia, Emetel Ecuador, CANTV of Venezuela, and Telecom Colombia.

"This is the first international South American cable to use SDH protocol, providing a capacity of 2.5 Gbps," according to Pioneer. The system will provide over 300,000 voice channels using multiplexing. The cable will be 4354 miles long, and it is estimated to cost $311 million.

Maya-1 will have landing points in Vero Beach, Florida; Cancun, Mexico; Puerto Cartes, Honduras; the Cayman islands; Colon, Panama; and Barranquilla, Colombia. This is a joint project of France Telecom and Telmex. The network will operate at 2.5 Gbps using three fiber pairs, and it will deploy underwater branching units. The network will also be connected to other undersea systems in the region. Alcatel and Tyco Submarine were named vendors for the project.

Pan-American Crossing, one of four networks planned by Global Crossing, hopes to establish the first direct path of connectivity for Latin American countries to U.S. and Asian markets without accruing additional transit fees. It will span 8000 km, connecting California, Mexico, and Panama with South American cables.

Plans for Atlantis-2 include linking Europe, West Africa, and South America. The configuration will include Portugal, the Canary Islands, Cape Verde, Senegal, Brazil, and Argentina, according to Pioneer. Landing parties include CPRM, Telefonica, Cable Verde Telecom, Sonatel, Embratel, and Telintar. In all the 12,800-km cable is expected to cost $320 million, according to Pioneer. Alcatel and Pirelli are supplying both cable and installation.

Brazil and Venezuela already have an extensive domestic shoreline system, and Argentina, Peru, and Chile plan similar systems, Pioneer reports. The Pan-American Cable System is expected to provide much needed capacity to the West Coast of South America.

Pioneer believes that the greatest prospects for follow-up cables in the region lie between South America and Europe, South America and the Caribbean, South and Central Americas, and South and North Americas. The Internet is expected to grow in South America, says Ruderman, with potentially more than 1 million users in Brazil and some 80,000 in both Argentina and Chile as of 1998. Another major growth market is cable television throughout the continent, and that uses a lot of fiber, says Ruderman. "Governments in South America are very enthusiastic because without telecom you can't bring in business," says Ruderman. Certainly that belief and improved conditions that will ensue in South America are destined to drive these networks to completion.

Section 6: An Enlightened Global Community

Telecommunications—and its primary conduit, fiber optics—is changing the world, and we are all included in the revolution effected by fax, E-mail, the Internet, electronic commerce, and other advanced communications services.

Bill Gates's comments about the information age and the information superhighway are generally correct although he and Andy Grove were more involved with creating content and how to push content with personal computers, leaving to others the details of how to move that content across telecommunications lines. While we are becoming more aware of the necessity of clean, bandwidth-rich transmission networks (Gates has even referred to fiber optics as the "asphalt of the global information superhighway") this awareness grew slowly, so that in some cases our carriers played catch-up as they were bombarded with data from such services as America On-Line.

According to Gates:

> The global information market will be huge and will combine all the various ways human goods, services, and ideas are exchanged.... On a practical level, this will give you broader choices about most things, including how you earn and invest, what you buy and how much you pay for it, who your friends are and how you spend your time with them, and where and how securely you and your family live (*The Road Ahead*, Viking Penguin, 1995, p. 6).

While Gates tends to view events in terms of personal growth, in many respects the societal impact will be far greater, and the Japanese, whose small nation was completely wired with fiber by 1983, understand this better than anyone perhaps. They were probably the first to recognize that productivity flourishes when communications are swiftly and clearly transmitted.

What impact does a society experience when it can suddenly conduct its business better on an individual basis? For one thing it gives rise to greater dependency on our own telecommunications lines and less dependency on a broader organizational PBX. As the Japanese believe, this may lead to a more decentralized form of society: Millions of commuters may not have to rush to Tokyo or Kyoto—or New York City or London—but instead work from their own home. In the case of Japan this obviously has important positive ramifications for an already-taxed infrastructure.

The Japanese see less moving around as generally a positive attribute, particularly for their elderly population. For example if an infirm patient's urine can be monitored by an off-site health facility from instruments in the patient's own bathroom, it may render transporting such patients to a facility for routine health checkups obsolete.

We may soon be able to scan a store's inventory from our own computer as the information age progresses.

In Chapter 15 we study in more detail a world where to some extent we become ubiquitous. It is an exciting new world made increasingly possible by a medium known as fiber optics that carries photons rather than electrons.

15

Instantaneous Global Communications

Fiber optics provides the potential for communicating with anyone in the world instantaneously, whether by telephone, e-mail, or facsimile. For example today we can send all of the Mother's Day telephone messages in the United States over one fiber simultaneously. In the 1980s it was a major challenge and success to have a laser commercially operating at 1 Gbps, or 1 billion bits per second. Such companies as Nortel Networks, Lucent, and NEC now say that they intend to deliver a terabit, or a trillion bits per second of information, by the millennium or thereabouts, something that was proven in the laboratory in 1996.

This is all part of what futurist George Gilder labels the law of the Telecosm:

> As communicators move up-spectrum, they can use bandwidth as a substitute for power, memory and switching....
>
> For more than a decade American companies have been laying optical fiber strands at a pace of some 4000 miles a day, for a total of more than 25 million strand miles.... Five years ago, the top 10% of U.S. homes and businesses were, on average, a thousand households away from a fiber node; now, they are a hundred households away.

A defining moment in the revolution occurred when Ciena introduced its 16-channel DWDM units, which multiplied by 16 the amount of capacity that had traditionally been used on a fiber system. A network that in its 20-year life used to feeding one laser per fiber suddenly could transmit up to 16 such information streams. Another bandwidth spurt occurred when Nortel Networks successfully began making 10-Gbps lasers, again increasing the bandwidth, although this time by fourfold.

Since then there were numerous bandwidth advertisements, as vendors announced commercial capabilities from 160 Gbps–1.3 Tbps over a single fiber using WDM. It is not always easy to understand who has real systems and who has vaporware, but the inexorable march of higher and higher speed networking continues.

Lucent, whose predecessors helped to found fiber optics, refers to networking as an r/evolution. The revolution involves such fiber carriers of the 1990s as WorldCom, Level 3, and IXC, which are building new fiber optics networks that incorporate DWDM and EDFAs. The evolutionary refers to fiber optic networks in the 1980s, which continue to offer faster and faster speeds and greater and greater service capabilities despite the limitations of an older technological base.

National pride is also involved, and it is a "very positive dynamic," because nations either want to lead or ensure that they do not lag too far behind, says Gates. He warns that such networks must be based on real-world needs in contrast to networks built by governments that may not grasp a rapidly changing technology (*The Road Ahead*, Viking, Penguin, 1995, p. 238).

For future direction perhaps the best place to look is the FSAN group, which represents the leading fiber optic vendors and carriers in the world who are developing close-in fiber optic systems. Siecor, and much of the fiber optics industry along with it, continues to wait for fiber to the home to become a reality. Says Siecor's Lawrence:

> If you map the classical or historical copper network, something like only 10% of the market is in that portion that is outside a mile of everyone's house while 90% of the network is in that last piece.

And even though fiber to the desk will be a major accomplishment, the amount of cable necessary to connect different floors of a business together is relatively small. According to Lawrence:

> However, the minute you try to go to people's own lines, cumulatively the volume is huge.... If and when fiber to the home becomes a reality, it is likely to be a pretty exponential increase and it can be of different products than we are making.... The cable likely will feature very low fiber counts, such as one or two fibers per cable.

Concerning future trends Lawrence continues to see growing demand for more fibers in a cable: "These might typically be deployed in metropolitan areas where they put rings around the city.... They will keep putting more and more fiber into these rings."

Trends to make optical fiber cable and ancillary equipment more user friendly will continue as fiber continues to be built deeper into the local exchange. Lawrence believes networking trends will continue to focus on making networks more optical, with as little electro-optical conversion as possible, although he acknowledges that Siecor is somewhat on the outside looking in:

> The trend is going to be there more and more to stay optical all the way through the system with as little electro-optic conversion as possible, and that can open up a whole new set of products.

The closer optical fiber cable comes into the network, the more hardware it requires. "Fiber to the desk will be very hardware intensive; the hardware will become a higher percentage of the overall network," says Lawrence.

The indisputable driver in all of this is the optical network. Photons work better, faster, and cheaper than electrons, so these will be the network of choice. Unfortunately it is not easy to move an entire architecture from electronics to photonics. Yet that is what a fully capable network requires, much to the consternation of those caught in the middle. According to Gilder:

> This means replacing nearly all the hundreds of billions of dollars' worth of switches, bridges, routers, converters, codecs, compressors, error correctors, and other devices, together with the trillions of lines of software code, that pervade the intelligent switching fabric of both telephone and computer networks.

Yet despite the pain, that revolution is inevitable: It will not occur at a pace determined by engineers, although their development of the optical switching fabric and plumbing for the optical network are critical elements. It will occur at a pace decreed by business and consumer demand, which is already increasing rapidly and unlike only a few years ago, is unstoppable. Rare is the individual today who complains of having too much bandwidth. (Lucent has even openly suggested the potential of 173 trillion bits over one fiber. Clearly that potential is enormous.)

Such technologists as Will Hicks and Rod Alferness envision the day when each one has his or her own wavelength, capable of transmitting a personalized stream of information embracing customized voice, video, and data requirements.

The genius of Kao, who founded fiber optics, again comes to mind. Kao, who characterizes fiber as "a zero cost to the infinite bandwidth zero loss medium," believes people will use tho bandwidth intelligently, noting that shock will set in when it continues to become cheaper and cheaper.

The result will inevitably be a world brought closer together, what Buckminster Fuller referred to as fellow citizens of planet earth. Such openness will create accountability on the part of leaders and nations, so that governmental errors are universally and immediately censored, a summons assistance, natural disaster, global business communications are instantaneous.

No doubt many things must occur before all the advantages of the global fiber optics superhighway materialize: Billions of dollars spent, life styles disrupted, technological advances, and reduced telecommunications equipment costs. But fiber optics is the ultimate solution, and the photonics network will effect global communications.

Index